Lamarck to Darwin:
Contributions to Evolutionary Biology
1809–1859

H. Lewis McKinney
Editor and Translator

Coronado Press

1971

Published by
Coronado Press
Box 3232
Lawrence, Kansas 66044

SBN 87291–019–9

First Published 1971
Reprinted, with corrections 1973
Third Printing 1975

Manufactured in the USA

CONTENTS

Introduction

While Wallace and Darwin ushered in the new era in evolutionary biology, they were by no means the first evolutionists, nor the first to describe the principle of natural selection. The crucial question is why did their precursors fail to convince the scientific community that species do, in fact, evolve? One possible answer is that some of these precursors – Wells, Matthew, and Blyth – had no such object in mind, although all of them to some degree touched on organic variation, evolution, and/or natural selection. Moreover, Matthew vociferously argued after the appearance in 1859 of the *Origin of Species* that the idea of natural selection was originally his; a chauvinistic American directed attention to the earlier work of Wells; and recently Loren Eiseley has claimed that Darwin misappropriated the idea of natural selection from Blyth. The fact that Darwin annotated his own personal copy of Blyth's article adds fuel to the fire. Obviously, a careful reading of the original sources is imperative if we are to understand the history of evolutionary biology well enough to evaluate the claims and counter claims of the various partisans.

Lamarck, Chambers and Wallace (1855) present different problems than do Matthew, Wells, and Blyth, for each argued vigorously and specifically for acceptance of the theory of evolution. Nevertheless, their arguments were overwhelmingly rejected, although Wallace's exceptional paper of 1855 was read carefully by Charles Lyell, Edward Blyth, and Charles Darwin. Despite the fact that the arguments of Lamarck and Chambers were unacceptable to many scientists, evolution could no longer be easily ignored after 1809 and especially after 1844. We find a typical response to evolutionists in Charles Lyell's *Principles of Geology* (1832 *et seq.*): species do vary, but only within predetermined limits. Both Wallace and Darwin reacted strongly to Point 4 of Lyell's "Recapitulation" of arguments regarding the reality of species in nature.

The following essays and excerpts are significant statements on organic variation, evolution, and natural selection up to and including Darwin's *Origin* of 1859. The selection from Lamarck's *Philosophie zoologique* (1809) is a new translation by me. Paula Gottdenker has checked the translation, although any errors remaining are my responsibility alone.

Since Lamarck has been so widely misunderstood and/or misrepresented, I have supplied the French terms for a number of important words. The word *circonstances*, for instance, is crucial in Lamarck's discussion. Generally, I have translated it as "circumstances," or "environmental circumstances" (or "conditions"), although simply "conditions," "conditions of life" (see Darwin's *Origin*, 1859, p. 43), or perhaps even "environment" might be just as satisfactory. Lamarck specifically means conditions of existence – such as habitation, exposure, climate, food, way (mode) of life, amount of heat, light, humidity, etc. – when he speaks of *circonstances* (see Lamarck, 1809, Vol. 1, pp. 223 and 225).

While Lamarck's presentation in 1809 no doubt was his most widely known discussion of evolution, he revised his ideas in 1815. I have translated only his four laws, although his entire discussion is fascinating and highly recommended. The French text and another English translation of these laws are available in Henry F. Osborn's *From the Greeks to Darwin* (1894 *et seq.*).

All other selections in this volume have been reset from the originals. Wells, Blyth (1835) and Wallace (1855) are complete. Matthew's work is from the Appendix to his book, sections B, C, (following) F. The Darwin-Wallace papers of 1858 are included, since they mark the beginning of Year One for modern evolutionary biology. The selection from Darwin's *Origin* (1859) provides an excellent summary of his arguments in his own words.

H. Lewis McKinney
Lawrence, Kansas
October 20, 1971

1809

J. B. P. A. Lamarck
The Influence of Circumstances

[J. B. P. A. Lamarck, "De l'influence des Circonstances sur les actions et les habitudes des Animaux, et de celle des actions et des habitudes de ces Corps vivans, comme causes qui modifient leur organisation et leurs parties," in *Philosophie Zoologique*. Paris: Dentu, 1809, Vol. 1, Ch. 7, pp. 221-235, 265-268]

I MUST EXPLAIN at this point what I mean by the statement: *Circumstances (circonstances) influence the form and organization of animals* – that is to say, when these circumstances become quite different they alter both the form and organization through proportional modifications during the course of time.

Certainly, if these statements were taken literally I could be accused of error; for, whatever environmental circumstances may do, they do not directly produce any modification whatsoever in the shape and organization of animals.

However, large changes in the environmental circumstances of animals bring about major changes in their needs, and such changes in turn necessarily bring about changes in their actions. Now, if their new needs become constant or especially long-lived, the animals then assume new *habits,* which are as long-lasting as the needs which produced them. This is easy to demonstrate, and indeed no explanation is required for it to be perceived.

It is, then, evident that a great change in the circumstances, having become constant for a race of animals, leads these animals to new habits.

Now, if new circumstances, having become permanent for a race of animals, have produced new *habits* in these animals – that is, have led them to new actions which have become habitual – the result will be the use of such parts that are preferred to others, and in certain cases, the total neglect [lack of use] of a part which has become useless.

None of this should be considered as hypothesis or personal opinion; on the contrary, these are truths which require only attention to and observation of the facts in order to be rendered evident.

By citing well-known substantiating facts, we shall see presently that, on the one hand, new needs – having made a part necessary – really have given birth to that part by virtue of repeated efforts; and that subsequently its constant use gradually has strengthened, developed, and finally enlarged it considerably. On the other hand, we shall see that in certain cases the new circumstances and new needs, having made a part completely useless, the total neglect of that part has resulted in its gradually ceasing to share in the development in which the other parts of the animal do share; that it gradually shrinks and diminishes; and that when there has been complete neglect for a long time, the part in question eventually disappears. All this is fact, for which I propose to give the most convincing proofs.

In plants, where there are no activities [such as those of animals] and consequently no *habits* per se, great alterations of environmental circumstances nonetheless lead to great differences in the development of their parts; so that these differences produce and develop some of them, while they reduce and cause the disappearance of others. But all this is brought about by the changes occurring in the nutrition of the plant, in its absorption and transpiration, in the quantity of heat [*calorique*], light, air, and humidity which it habitually receives; finally, in the superiority that certain of the different vital movements may assume over others.

Among individuals of the same species, some of which are constantly well nourished and in circumstances favorable to their development, while others live in exactly reversed circumstances, a difference arises in the condition of these individuals which gradually becomes very noticeable. How many examples I might cite in animals and plants which would confirm the grounds for this view! Now, if the circumstances remain the same, making habitual and constant the condition of individuals poorly nourished, injured or sickly, their internal organization eventually is modified, and reproduction between them preserves the acquired modifications, giving rise ultimately to a race very distinct from those found in circumstances which are always favorable to their development.

A very dry spring causes prairie grasses to grow but slightly, and they remain meagre and puny, flowering and fruiting without growing very much. A spring interspresed with both warm and rainy days makes these same grasses grow vigorously, and the hay harvest is then excellent. However, if anything perpetuates unfavorable circumstances for these plants, a proportional change occurs: first in their appearance and general condition, and finally in several particulars of their characteristics.

For example, suppose some seeds of any of the grasses in question were transported to a higher elevation on dry, barren and stony terrain, exposed to the winds, and still were able to germinate. The plant which could survive in such a place being always poorly nourished, and its offspring – continuing to exist in such adverse circumstances – would become a race very much different from the one which lives in the field, and yet it originated from that same field. The individuals of this new race will have small, meagre parts, and certain of their organs, having developed more than others, will present unusual shapes.

Those who have observed much and have consulted large collections have been convinced that as changes occur in the circumstances – habitation, exposure, climate, food, mode of life, etc. – the characteristics of size, shape, proportion of the parts, color, consistency, agility and dexterity in the animals change proportionally.

What nature accomplishes over a long period of time, we do every day by changing suddenly, in respect to a living plant, the circumstances in which it and all the individuals of its species are situated.

Every botanist knows that the plants taken from their birthplaces and placed in gardens for cultivation gradually undergo changes which finally make them unrecognizable. Many plants of a hairy nature become hairless, or nearly so; many plants which used to creep and trail on the ground become erect. Others lose their thorns or spines; still others, whose stems were woody and perennial in the hot climates where they lived, become herbaceous in our climate, and some of them become annuals. Finally, the dimensions of their parts themselves undergo very considerable changes. These effects of the changes of circumstances are so well known that botanists do not at all like to describe garden plants, unless they have come only recently under cultivation.

Was it not man who brought cultivated wheat *(Triticum sativum)* to its present state? Let someone tell me what country has a similar plant living in a natural state – that is to say, one which is there not as a result of cultivation in some neighboring area.

Where in nature do we find our cabbage, lettuce, etc., in the same condition as we find them in our kitchen gardens? Is not the case the same with a number of animals which have been changed or considerably modified by domestication?

How many different races of domesticated fowl and pigeons have we obtained by raising them in diverse circumstances and in different countries, and which we now would vainly seek to find in nature!

Those which are least changed, doubtless because their domestication is more recent and because they do not live in a climate foreign to them, nevertheless display great differences in some of their parts. These differences were produced by the habits which we made them adopt. Thus our ducks and domestic geese are of the same type as the wild ducks and geese; but ours have lost the ability to soar into the high regions of the air and to fly over extensive areas; finally, a definite change has been effected in the condition of their parts, compared to those animals from which they have descended.

Who does not know that if we raise a bird of our climate in a cage and it lives there five or six years in succession before being placed back in nature – i.e., set free – it is then unable to fly like its fellow creatures who have always been free. The slight change of circumstances for this individual has only diminished its ability to fly, and doubtless has not produced any change in the shape of its parts. But if a long series of generations of individuals of the same race had been kept in captivity for a considerable time, there is no doubt that the form of the parts of these individuals would gradually experience notable changes. For a much stronger reason, if instead of a simple continuous captivity, this environmental condition had been simultaneously accompanied by a change to a very different climate, and if these individuals had by degrees been habituated to other kinds of food and to other actions to obtain it, these environmental conditions would, when combined and permanently established, insensibly form a new and special race.

Where in nature do we now find this multitude of races of *dogs* which actually exist as a result of the domestication to which we have reduced these animals? Where do we find those bulldogs, greyhounds, water spaniels, spaniels, lap dogs, etc., etc., races which present greater differences among themselves than those which we call specific among animals of the same genus when living in a wild state.

Doubtless, a single, primitive race very near the wolf, if it is indeed not the true [original?] type, has been at some period domesticated by man. That race, all the individuals of which were alike, has been gradually dispersed by man into different countries with different climates; and after a time these same individuals, which have been subjected to the influences of their habitats and diverse habits forced on them in each country, have undergone remarkable changes and have formed different special races. Now, one who travels very great distances for commerce or interests of another kind brings various races of dogs – which have originated in very remote countries – into densely populated places, such as a large capital; then the crossing of these races of dogs by breeding has given rise successively to all those we now know.

As far as plants are concerned, the following fact proves how the change of some important environmental factor brings about a change in the parts of these living bodies.

So long as the *Ranunculus aquatilis* is submerged in the water, all its leaves are finely divided into hair-thin divisions; but when the stalks of this plant reach the surface of the water, the leaves which develop in the air are larger, rounded, and simply lobed. If several feet [*pieds*] of the same plant succeed in growing in a soil that is only damp, without being submerged, their stalks will be short, and none of their leaves will be divided into hair-fine divisions; this gives rise to *Ranunculus hederaceus*, which botanists regard as a separate species.

There is no doubt in regard to animals that important changes in circumstances in which they are accustomed to live produce similar changes in their parts; but here the transformations occur more slowly than in plants and consequently are less perceptible, and their cause less easily recognizable.

As to the environmental conditions which are so powerful in modifying the organs of living bodies, doubtless the most influential are the diversity of the environments [*milieux*] in which they live; but there are also many others which have considerable influence in the production of the effects in question.

It is known that different localities vary in nature and kind because of their position, arrangement, and their climate, as is easily seen in passing through different places distinguished by special features. There is, therefore, a cause of variation for animals and plants living in these diverse places. But what we do not know, and indeed what we generally refuse to believe, is that each place itself changes with time in regard to exposure, climate, nature and quality, although with such extreme slowness, as far as our own existence is concerned, that we attribute perfect *stability* to it.

Now, in both cases, these changed localities proportionately alter the circumstances relative to the organisms that live there, and this in turn produces other effects on those same organisms.

Thus we realize that if there are extremes in these changes, there are also mere shades of differences which, so to speak, form intermediate steps to fill the interval. Consequently, there are also slight gradations in the differences which distinguish what we call *species.*

It is evident then that the entire surface of the earth offers, in the nature and arrangement of the materials which occupy its different regions, a diversity of environmental conditions which everywhere impinge on that of the forms and parts of animals, independently of the particular diversity which necessarily results from the development of each animal's organizational make-up.

In each locality where animals are able to live, the environmental conditions which establish a particular pattern there may remain unchanged for a long time; [even then] they change so very slowly that man cannot observe them directly. He is forced to consult monuments [*monumens*] in order to recognize that in each place the order of things he discovers has not always been the same, and to realize that it will change still more.

The races of animals, as long as they live in each of these locations, should therefore preserve their habits for an equally long time: hence the seeming immutability of the races which we call *species.* This constancy has instilled in us the idea that these races are as ancient as nature.

But in the different habitable spots on the surface of the earth, the nature and situation of the places and climates constitute – for animals as well as plants – *different environmental conditions,* differing in all kinds of degrees. The animals living in those different places must then differ from each other – not only because of the organizational make-up of each race, but also because of habits that the individuals of each race are forced to adopt; thus, while traveling over large portions of the earth's surface, the observant naturalist sees that commensurate with a significant change in environmental conditions, there will always be a proportional change in the characteristics of species.

Now, the proper sequence of things to consider in all this is as follows:

1. Every fairly considerable and continuing change in the circumstances of any race of animals brings about a real change in their needs [*besoins*].

2. Every change in the needs of animals necessitates different actions to satisfy the new needs and, consequently, different habits.

3. Every new need, necessitating new actions to satisfy it, requires of the animal that it either [a] use certain parts more frequently than it did before, thereby considerably developing and enlarging them, or [b] use new parts which their [new] needs have imperceptibly developed in them, by virtue of the operations of their own inner sense [*par des efforts de son sentiment intérieur*]. I shall prove this shortly by known facts.

Therefore, to obtain a knowledge of the true causes of so many diverse forms and so many different habits of which [commonly] known animals offer us examples, we must consider that the infinitely diversified, but always slowly changing circumstances in which the animals are successively found, have introduced each of them to new needs and, of necessity, to changes in their habits. Now, this cannot be denied once it is recognized. We shall easily perceive how new needs may have been satisfied and new habits adopted, if we pay attention to the following two laws of nature, which observation always verifies:

FIRST LAW

*In every animal which has not yet passed beyond the limit [*terme*] of its development, the more frequent and sustained use of any organ gradually strengthens, develops and enlarges that organ, and gives it a strength proportional to the length of time it has been used; while the constant disuse of such an organ imperceptibly weakens and deteriorates it, progressively diminishing its faculties, until it finally disappears.*

SECOND LAW

Everything which nature has caused individuals to acquire or lose as a result of the influence of environmental conditions to which their race has been exposed over a long period of time – and consequently, as a result of the effects caused either by the extended use (or disuse) of a particular organ – [all this] is conveyed by generation to new individuals descending therefrom, provided that the changes acquired are common to both sexes, or to those which have produced the new individuals.

These are the two fundamental truths which can be ignored only by those who have never observed or followed the operations of nature, or by those who have allowed themselves to be carried away by the mistaken notion with which I now propose to deal....*

◊

[Page 265] .... We need not adduce further examples and may now proceed to the [main] point of the discussion.

The fact is that the diverse animals – each according to their own genus and species – have special habits, and an organization which is always perfectly in harmony with those habits.

From the consideration of this fact, it might appear that we are free to admit either one or the other of the two following conclusions, – and that neither of them can be proven.

Conclusion accepted until now: Nature (or the Author of Nature), in creating animals, has foreseen all the possible kinds of circumstances [and conditions] in which they would have to live, and has given to each species a stable organization, as well as a form which is fixed and invariable in its parts. These force each species to live in the places and climates where we find them, and to preserve the habits which are characteristic of it.

My personal conclusion: Nature, in successively producing all species of animals, beginning with the most imperfect or the simplest, and ending her work with the most perfect, has caused their organization gradually to become more complex. These animals are generally distributed throughout all the habitable regions of the earth, each species coming under the influence of the environmental conditions in which it finds itself, acquiring the habits by which we know them and the modifications in their parts which observation reveals to us.

The first of these two conclusions is that which has been held up to the present, that is to say, it is accepted by almost everyone. It assumes in each animal an unchanging organization and parts which have never varied, and which never will vary. It furthermore assumes that the

---

*Omitted here are pages 236 through 264

environmental conditions of the regions inhabited by each species of animal will never vary; for if they should vary, these same animals could not go on living there, nor could they possibly find a new habitat [like the old] to which they could move.

The second conclusion is my own. It assumes that through the influence of environmental conditions on the habits, and the effect of [the new] habits on the condition of the parts – and on the make-up [of the animal itself] – each animal may undergo modifications (perhaps considerable ones) in its parts and organization. Such modifications give rise to the condition in which we find all animals [today].

To demonstrate that the second conclusion is without foundation, it must first be proved that each point on the surface of the earth never varies in its nature, exposure, high or low elevation, climate, etc., etc.; and it must then be proved that no part of animals undergoes – even after a long period of time – any modification caused by a change of environmental circumstances, and by the necessity which forces them into another kind of life and activity than the one to which they have been accustomed.

Now, if a single fact establishes that an animal which has been under domestication for a long time differs from a wild species from which it has descended, and if in such a species under domestication we discover a great difference of structure between those subjected to such habit and those forced into different habits, – then it will be certain that the first conclusion does not conform to the laws of nature, and that on the contrary, the second conclusion is perfectly in accord with them.

Everything concurs then to prove my assertion: namely, that it is not the form – either of the body or the parts – which gives rise to the habits and manner of life of animals. On the contrary, it is the habits, the manner of life, and all the other environmental circumstances which have, during the course of time, shaped the form of the body and the parts of animals. With new forms, new faculties have been acquired, and gradually nature has been able to shape the animals as we see them now.

Can there be a more important consideration in natural history, and [one] to which we should devote more attention than that which I have just set forth?

•

1. La vie, par ses propres forces, tend continuellement à accroître le volume de tout corps qui la possède, et à étendre les dimensions de ses parties, jusqu'à un terme qu'elle amène elle-même.

2. La production d'un nouvel organe dans un corps animal, résulte d'un nouveau besoin survenu qui continue de se faire sentir, et d'un nouveau mouvement que ce besoin fait naître et entretient.

3. Le développement des organes et leur force d'action sont constamment en raison de l'emploi de ces organes.

4. Tout ce qui a été acquis, tracé ou changé, dans l'organisation des individus, pendant le cours de leur vie, est conservé par la génération, et transmis aux nouveaux individus qui proviennent de ceux qui ont éprouvé ces changements.

1815

J. B. P. A. Lamarck
Four Evolutionary Laws

[J. B. P. A. Lamarck, *Histoire Naturelle des Animaux sans Vertèbres.* 7 Vols. Paris, 1815-1818. The translation given here is from the French text as reprinted (with his own English translation) in Henry Fairfield Osborn's *From the Greeks to Darwin.* London and New York: The Macmillan Company, 1908, pp. 166-167]

1. Life by its own forces tends continually to increase the volume of every body that possesses it, as well as to increase the size of all the parts of the body up to a limit which it imposes upon itself.

2. The production of a new organ in the body of an animal results from a new need arriving unexpectedly and continuing to be felt, and from the new movement which this need initiates and maintains.

3. The development of organs and their power of action are always proportional with the use of these organs.

4. Everything that has been acquired, delineated, or altered in the organization of individuals during their life is preserved by generation and transmitted to new individuals proceeding from those which have undergone these changes.

1818

William Charles Wells
An Account of a Female of the White Race of Mankind

[William Charles Wells, "An Account of a Female of the White Race of Mankind, Part of Whose Skin Resembles that of a Negro; With Some Observations on the Causes of the Differences in Colour and Form between the White and Negro Races of Men," in *Two Essays: One upon Single Vision with Two Eyes; The Other on Dew.* London: Archibald Constable and Co., 1818, pp. 425-439]

INSTANCES of the absence of the black colour, in the whole or part of the skin in persons of the negro race, are not very uncommon; but there is, I believe, no one upon record of an individual of the white race having any part of the body, covered with a skin similar to that of a negro. The following account, therefore, of such an instance, will, perhaps, be acceptable to the philosophical public. I have been enabled to form it by the permission of Dr. Turner, one of my colleagues at St. Thomas's Hospital, into which the person, whose case I am to describe, was lately admitted by him, on account of some bodily ailment.

Hannah West, the subject of this account, was born in a village in Sussex, about three miles distant from the sea, and is now in the twenty-third year of her age. Both of her parents were natives of the same county. Her father was a footman in a gentleman's family, and died while she was very young. She cannot, therefore, remember his appearance; but she has never heard, that it was in any way extraordinary. Her mother is still alive, and has black hair and hazel eyes, but a fair skin, without any stain or mark upon it. West was the only child of her father; but her mother, having married a second time, has had eleven other children. Nine of these are living, all of whom are without any blackness of the skin. Her mother, she says, received a fright, while pregnant with her, by accidentally treading on a live lobster; and to this was attributed the blackness of part of her skin, which was observed at her birth.

West is somewhat above the middle size, is rather of a full habit, and till she came to London from Sussex, which was about four months ago, always enjoyed very good health. The hair of her head is of a light brown colour, and is very soft; her eyes of a faint blue; her nose prominent and a little aquiline; her lips thin; the skin of the greater portion of the uncovered parts of her body very white; in short, her appearance is in every respect, except the one which has been mentioned, that of a very fair female of the white race of mankind.

The parts covered by the black skin are, the left shoulder, arm, fore-arm, and hand. All these parts, however, are not universally black; for on the outside of the fore-arm, a little below the elbow, a stripe of white skin commences, about two inches in breadth, and differing in no circumstance from the skin of the other arm, which, proceeding upwards, gradually bends under the arm, and at the arm-pit joins with the white skin of the trunk of the body. The black skin, wherever it is contiguous to the white, terminates rather abruptly, so that its boundary may always be distinctly traced.

The colour of the black skin is not everywhere uniformly dark. Thus, the skin of the back of the hand, and of the wrist, is marked by fine lines of a reddish black, which cross one another at right angles, while the small rectangular spaces bounded by these lines are entirely black. Part of the cuticle of the hand having been removed by exciting a blister, the reddish lines were found to be the summits of very thin folds of the true skin, which were raised above its general level, and were less thickly covered with the black *rete mucosum* than the more depressed parts. Their reddish colour was, no doubt, occasioned by the external air, as the skin of the other hand was red from that cause. All the other parts of the black skin are fully as dark, as I found on making the comparison, as the corresponding parts of a dark negro, and are much darker than those of many negroes. One part, indeed, of her skin is considerably darker than the corresponding part in any negro whom I have seen; for the palm of her hand and inside of her fingers are black, whereas these parts in a negro are only of a tawny hue.

A considerable part of the black skin is as smooth to the touch, as the skin of the white arm; but the cuticular lines in the black arm, appeared everywhere stronger to the sight, than similar lines in the arm of a black man, whose skin I examined at the same time. In the greater part, however, of West's black skin, those lines sink deeper beneath its general surface, than the lines of any other human skin that I have seen, which was not evidently diseased. These depressions are extremely narrow, and proceed chiefly in one direction, obliquely upwards and inwards from the outer part of the arm. On removing a small portion of the cuticle, they were found to be occasioned by the sinking down of that membrane between very narrow and slightly elevated folds of the true skin, nearly contiguous to one another, which held the direction mentioned.

A great part of the black shoulder exhibits a singular appearance; for, near to the back bone, the skin, over an extent of six inches in length and two in breadth, resembles a thick coat of pitch, or black paint, which by drying had split into a great number of small square portions. The fissures in the skin are about a line in depth. Mr. James Wilson, teacher of anatomy, and fellow of the Royal Society, who saw this person once along with me, pulled away a little of this black matter, upon which several narrow processes of the skin, perpendicular to the plane of the part, became visible.

Winslow says, that the cuticle of a negro is black, and that the contrary supposition arose from its tenuity and transparency, in like manner as a thin film of black horn appears almost colourless. I have found by my own observations, that this opinion of Winslow is just; and I found also, that the cuticle of West's black skin is likewise dark. I may add, that the nails of her black fingers are darker than those of the white, and darker also than those of a negro's hand.

Sir Everard Home, who likewise saw this person once along with me, thought that the black arm smelt more strongly than the white. I made the experiment immediately after him, and thought so too. But on repeating it several times with more attention, I could perceive no difference. It seems to me, indeed, from a similar experiment made on the arm of a dark negro, whose appearance did not lead me to suppose, that he had been very careful with respect to the cleanliness of his person, either that all negroes do not possess a strong smell, or that this does not proceed from all parts of their skin, since I could perceive no difference between the odour of his arm, and that of the white arm of West.

On the black fore-arm are about a dozen small hard substances, the largest of which are of the size of a common pea. Some of them are very black; others are less black, and one or two are of a reddish black colour. I thought, at first, that they consisted of thickened cuticle, but

I found afterwards, that they readily bled upon being punctured with a needle.

The upper and outer part of the black arm has a number of very black hairs growing from it, some of which are three quarters of an inch long. The inner part of the arm, which is equally black, is free from hairs.

The black arm is as firm to the touch, and as fleshy as the white; and according to the young woman's own report, there is no difference in their strength or feelings of any kind.

The last circumstance which I shall mention concerning her is, that no change has taken place within her remembrance, either in the degree or extent of the blackness of her skin.

Two inferences may, I think, be made from what has been related respecting Hannah West.

The first is, that the blackness of the skin in negroes is no proof of their forming a different species of men from the white race.

When a white man is much exposed to the action of the sun, his skin becomes more or less brown, and as the intensity of this colour, after equal degrees of exposure, is generally proportional to the heat of the climate, it has hence been supposed, that the colour of negroes is derived from a very great degree of the same cause. But this conclusion seems to me very faulty. For, setting aside that a white man, rendered brown by the sun's rays, begets as white children as those of another of the same race, the colour of whose skin had never been altered, it appears to me probable, from observations lately made on two negroes, that the action of the sun tends rather to diminish than augment the colour of their race. Both of those persons were born in European settlements, and had been accustomed to have their bodies clothed, yet, in both, the trunk, arms, and lower extremities, were considerably darker than the face, and in one, were somewhat darker than the hands. But admitting this observation to be of no force, still it must be granted, in consequence of what has been said upon the state of part of West's skin, —that great heat is not indispensably necessary to render the human colour black; which is the second conclusion to be drawn from the account which has been given of her.

On considering the difference of colour between Europeans and Africans, a view has occurred to me of this subject, which has not been given by any author, whose works have fallen into my hands. I shall, therefore, venture to mention it here, though at the hazard of its being thought rather fanciful than just.

There is no circumstance, perhaps, in which these two races differ so much, as in their capacity to bear, with impunity, the action of the causes of many diseases. The fatality to Europeans of the climate of the middle parts of Africa, which are, however, inhabited by negroes without injury to their health, is well known. Let it then be supposed, that any number of Europeans were to be sent to that country, and that they were to subsist themselves by their bodily labour; it seems certain, that the whole colony would soon become extinct. On the other hand, the greater liability of negroes in Europe to be attacked with fatal diseases is equally well established. If, therefore, a colony of the former race were brought to Europe, and forced to labour in the open air for their subsistence, many of them would quickly die, and the remainder, from their inability to make great bodily exertions in cold weather, and their being frequently diseased, would be prevented from working an equal number of days in the year with the whites. The consequence would be, that without taking farther into account the unfriendliness of the climate to them, their gains would be inadequate to the maintenance of themselves and their families. They would thence become feeble, and be rendered still more incapable of supporting life by their labour. In the mean time, their children would die from want, or diseases induced by deficient or improper nourishment, and in this way, a colony of the negro race in a cold country would quickly cease to exist.

This difference in the capacity of the two races to resist the operation of the causes of many diseases, I assume as a fact, though I am utterly unable to explain it. I do not, however, suppose, that their different susceptibility of diseases depends, properly, on their difference of colour. On the contrary, I think it probable, that this is only a sign of some difference in them, which, though strongly manifested by its effects in life, is yet too subtle to be discovered by an anatomist after death; in like manner as a human body, which is incapable of receiving the small-pox, differs in no observable thing from another, which is still liable to be affected with that disease.

Regarding then as certain, that the negro race are better fitted to resist the attacks of the diseases of hot climates than the white, it is reasonable to infer, that those, who only approach the black race, will be likewise better fitted to do so, than others who are entirely white.

This is, in fact, found to be true, with regard to the mixture of the two races; since mulattoes are much more healthy in hot climates than whites. But amongst men, as well as among other animals, varieties of a greater or less magnitude are constantly occurring. In a civilized country, which has been long peopled, those varieties, for the most part, quickly disappear, from the intermarriages of different families. Thus, if a very tall man be produced, he very commonly marries a woman much less [tall] than himself, and their progeny scarcely differs in size from their countrymen. In districts, however, of very small extent, and having little intercourse with other countries, an accidental difference in the appearance of the inhabitants will often descend to their late posterity. The clan of the Macras, for instance, possess both sides of Loch-Duich in Scotland; but those who inhabit one side of the loch are called the black Macras, and the others the white, from a difference which has always been observed in their complexions. Again, those who attend to the improvement of domestic animals, when they find individuals possessing, in a greater degree than common, the qualities they desire, couple a male and female of these together, then take the best of their offspring as a new stock, and in this way proceed, till they approach as near the point in view, as the nature of things will permit. But, what is here done by art, seems to be done, with equal efficacy, though more slowly, by nature, in the formation of varieties of mankind, fitted for the country which they inhabit. Of the accidental varieties of man, which would occur among the first few and scattered inhabitants of the middle regions of Africa, some one would be better fitted than the others to bear the diseases of the country. This race would consequently multiply, while the others would decrease, not only from their inability to sustain the attacks of disease, but from their incapacity of contending with their more vigorous neighbours. The colour of this vigorous race I take for granted, from what has been already said, would be dark. But the same disposition to form varieties still existing, a darker and a darker race would in the course of time occur, and as the darkest would be the best fitted for the climate, this would at length become the most prevalent, if not the only race, in the particular country in which it had originated.

In like manner, that part of the original stock of the human race, which proceeded to the colder regions of the earth, would in process of time become white, if they were not originally so, from persons of this colour being better fitted to resist the diseases of such climates, than others of a dark skin.

The cause which I have stated, as likely to have influence on the colour of the human race, would necessarily operate chiefly during its infancy, when a few wandering savages, from ignorance and improvidence, must have found it difficult to subsist throughout the various seasons of the year, even in countries the most favourable to their health. But, when men have acquired the knowledge of agriculture, and other arts, and in consequence adopt a more refined mode of life, it has been found, that an adherence to their ancient customs and practices will preserve them long as a distinct race from the original inhabitants of the country to which they had emigrated. Examples of this kind are frequent in the islands in the eastern seas in the torrid zone, where the inhabitants of the sea-coast, evidently strangers, are in some degree polished, and of a brown colour, while the ancient natives, who live in the interior parts, are savage and black. Similar facts occur in respect to other species of animals. It seems certain, for instance, that fine woolled sheep, like the Spanish, never both arose and sustained their breed in the northern parts of Europe; yet, by care, this feeble race, after being formed in Spain, has been propagated and preserved in very cold countries. Thus the late Mr. Dryander, the learned librarian of the Royal Society, informed me, that the breed of fine woolled Spanish sheep had been kept perfect in Sweden during a very long term of years, I think he said a century. If, then, my memory be accurate upon this point, we have here an example of a variety of animals, much more liable to be affected by external circumstances than the human race, being preserved without change, in a country very different from their own, by assimilating their new state as much as possible to their old, during at least fifty generations, that is, during a period equivalent to 1500 years in the history of man.

Hitherto, while speaking of the external appearance of negroes, I have taken notice only of their colour. I shall now say a few words upon their woolly hair, and, according to our notions of beauty, the deformity of their features.

There are several facts which seem to show, that these circumstances are somehow connected with their low state of civilization.

First; the black inhabitants of the Indian Peninsula within the Ganges, who, compared with the African negroes, may be regarded as a polished people, have hair and features much less dissimilar to the European.

Secondly; Woolly heads, and deformed features, appear again, as we proceed further to the east, among the savage inhabitants of New

Guinea, and the adjacent islands, at the distance of nearly half of the circumference of our globe from Africa, and consequently without the smallest probability of any communication having ever existed between the two countries.

Lastly; it appears probable from the reliques of ancient art, that the early inhabitants of Egypt were of the negro race. If, then, the negroes of Africa were ever to be civilized, their woolly hair and deformed features would, perhaps, in a long series of years, like those of the Egyptians, be changed. On the other hand, their present external appearance may possibly be regarded not only as a sign, but as a cause of their degraded condition, by preventing, in some unknown way, the proper development of their mental faculties; for the African negroes have in all ages been slaves; and the negroes in the eastern seas are in no instance, I believe, masters of their handsomer neighbours, but are in many places in entire subjection to them, though the latter be frequently less numerous.

It will no doubt be objected to what I have advanced respecting the difference of colour between Europeans and Africans, that the Indian inhabitants of the greater part of the immense continent of America have skins nearly of one hue. Plausible reasons may, I think, be given for this fact, consistently with what has been said upon the colour of the two former races; but I forbear trespassing any longer upon the time of the reader, in discussing a subject which admits only of conjectural reasoning.

1831

Patrick Matthew
Notes on Natural Selection

[Patrick Matthew, *On Naval Timber and Arboriculture; With Critical Notes on Authors Who Have Recently Treated the Subject of Planting.* London: Longman, Rees, Orme, Brown, and Greene; Edinburgh: Adam Smith, 1831, Appendix, Notes B, C, following F, pp. 364-375, 381-388]

Note B

THERE IS a law universal in nature, tending to render every reproductive being the best possibly suited to its condition that its kind, or that organized matter, is susceptible of, which appears intended to model the physical and mental or instinctive powers, to their highest perfection, and to continue them so. This law sustains the lion in his strength, the hare in her swiftness, and the fox in his wiles. As Nature, in all her modifications of life, has a power of increase far beyond what is needed to supply the place of what falls by Time's decay, those individuals who possess not the requisite strength, swiftness, hardihood, or cunning, fall prematurely without reproducing – either a prey to their natural devourers, or sinking under disease, generally induced by want of nourishment, their place being occupied by the more perfect of their own kind, who are pressing on the means of subsistence. The law of entail, necessary to hereditary nobility, is an outrage on this law of nature which she will not pass unavenged – a law which has the most debasing influence upon the energies of a people, and will sooner or later lead to general subversion, more especially when the executive of a country remains for a considerable time efficient, and no effort is needed on the part of the nobility to protect their own, or no war to draw forth or preserve their powers by exertion. It is all very well, when, in stormy times, the baron has every faculty trained to its utmost ability in keeping his proud crest aloft. How far hereditary nobility, under effective government, has operated to retard "the march of intellect," and deteriorate the species in modern Europe, is an interesting and important question. We have seen it play its part in France; we see exhibition of its influence throughout the Iberian peninsula, to the utmost degradation of its victims. It has rendered the Italian peninsula, with its islands, a blank in the political map of Europe. Let the panegyrists of hereditary nobility, primogeniture, and entail, say what these countries might not have been but for the baneful influence of this unnatural custom. It is an eastern proverb, that no king is many removes from a shepherd. Most conquerors and founders of dynasties have followed the plough or the flock. Nobility, to be in the highest perfection, like the finer varieties of fruits, independent of having its vigour excited by regular married alliance with wilder stocks, would require stated complete renovation, by selection anew, from among the purest crab. In some places, this renovation would not be so soon requisite as in others, and, judging from facts, we would instance Britain as perhaps the soil where nobility will continue the longest untainted. As we advance nearer to the equator, renovation becomes sooner necessary, excepting at high elevation – in many places, every third generation, at least with the Caucasian breed, although the finest stocks be regularly imported. This renovation is required as well physically as morally.

It is chiefly in regard to the interval of time between the period of necessary feudal authority, and that when the body of the population having acquired the power of self-government from the spread of knowledge, claim a community of rights, that we have adverted to the use of war. The manufacturer, the merchant, the sailor, the capitalist, whose mind is not corrupted by the indolence induced under the law of entail, are too much occupied to require any stimulant beyond what the game in the wide field of commercial adventure affords. A great change in the circumstances of man is obviously at hand. In the first step beyond the condition of the wandering savage, while the lower classes from ignorance remained as helpless children, mankind naturally fell into clans under paternal or feudal government; but as children, when grown up to maturity, with the necessity for protection, lose the subordination to parental authority, so the great mass of the present population requiring no guidance from a particular class of feudal lords, will not continue to tolerate any hereditary claims of authority of one portion of the population over their fellow-men; nor any laws to keep up rank and wealth corresponding to this exclusive power. – It would be *wisdom* in the noblesse of Europe to abolish every claim or law which serves to point them out a separate class, and, as quickly as possible, to merge themselves into the mass of the population. It is a law manifest in nature, that when the use of any thing is past, its existence is no longer kept up.

Although the necessity for the existence of feudal lords is past, yet the same does not hold in respect to a hereditary head or King; and the stability of this head of the government will, in no way, be lessened by such a change. In the present state of European society, perhaps no other rule can be so mild and efficient as that of a liberal benevolent monarch, assisted by a popular representative Parliament. The poorest man looks up to his king as his own, with affection and pride, and considers him a protector; while he only regards the antiquated feudal lord with contempt. The influence of a respected hereditary family, as head of a country, is also of great utility in forming a principle of union to the different members, and in giving unity and stability to the government.

In respect to our own great landholders themselves, we would ask, where is there that unnatural parent – that miserable victim of hereditary pride – who does not desire to see his domains equally divided among his own children? The high paid sinecures in church and state will not much longer be a great motive for keeping up a powerful family head, whose influence may burthen their fellow-citizens with the younger branches. Besides, when a portion of land is so large, that the owner cannot have an individual acquaintance and associations with every stream, and bush, and rock, and knoll, the deep enjoyment which the smaller native proprietor would have in the peculiar features, is not called forth, and is lost to man. The abolition of the law of entail and primogeniture, will, in the present state of civilization, not only add to the happiness of the proprietor, heighten morality, and give much greater stability to the social order, but will also give a general stimulus to industry and improvement, increasing the comforts and elevating the condition of the operative class.

In the new state of things which is near at hand, the proprietor and the mercantile class will amalgamise, – employment in useful occupations will not continue to be held in scorn, – the merchant and manufacturer will no longer be barely tolerated to exist, harassed at every turn by imposts and the interference of petty tyrants; – Government, instead of forming an engine of oppression, being simplified and based on morality and justice, will become a cheap and efficient protection to person and property; and the necessary taxation being levied from property alone, every individual will purchase in the cheapest market, and sell the produce of his industry in the dearest. This period might, perhaps, be accelerated throughout Europe, did the merchants and capitalists only know their own strength. Let them, as citizens of the world, hold annual congress in some central place, and deliberate on the interests of man, which is their own, and throw the whole of their influence to support liberal and just governments, and to repress slavery, crime, bigotry – tyranny in all shapes. A Rothschild might earn an unstained fame, as great as yet has been attained by man, by organizing such a power, and presiding at its councils.

Note C

The influence of long continued impression, constituting instinct or habit of breed, is a curious phenomenon in the animal economy. Our population in the eastern maritime districts of Britain, descended principally from the Scandinavian rover, though devoted for a time to agricultural or mechanical occupation, betake themselves, when opportunity offers, to their old element, the ocean,* and launch out upon the "wintry wave" with much of the same home-felt composure as does the white polar bear. They roam over every sea and every shore, from Behring's Straits to Magellan's, with as little solicitude as the Kelt over his own misty hill, overcoming, in endurance, the native of the torrid zone under his vertical sun, and the native of the frigid among his polar snows.

To what may we ascribe the superiority of this portion of the Caucasian breed, – may it arise in part from its repeated change of place under favourable circumstances? Other races have migrated, but

* The habit of breed is apparent in many places of the world. Where a fine river washes the walls of some of the internal towns of France, scarce a boat is to be seen, except the long tract-boats employed in the conveyance of fire-wood— nobody thinks of sailing for pleasure. The Esquimaux, and the Red Indian of North America, inhabiting the same country, shew an entirely distinct habit of breed. The Black and the Copper-coloured native of the Australian Islands, are equally opposed in instinctive habit.

not like this, always as conqueror. The Jew has been a stroller in his time; but he has improved more in mental acumen and cunning – not so much in heroism and personal qualities: his proscribed condition will account for this. The Caucasian in its progress, will also have mingled slightly, and, judging from analogy, perhaps advantageously, with the finer portion of those whom it has overwhelmed. This breed, by its wide move across the Atlantic, does not seem at all to have lost vigour, and retains the nautical and roving instinct unimpaired, although the American climate is certainly inferior to the European. It is there rapidly moving west, and may soon have described one of the earth's circles. A change of seed, that is, a change of place, within certain limits of latitude, is well known to be indispensable to the more sturdy growth and health of many cultivated vegetables; it is probable that this also holds true of the human race. There are few countries where the old breed has not again and again sunk before the vigour of new immigration; we even see the worn out breed, chased from their homes to new locations, return, after a time, superior to their former vanquishers, or gradually work their way back in peace, by superior subsisting power: this is visible in France, where the aboriginal sallow Kelt, distinguished by high satyr-like feature, deep-placed sparkling brown or grey eye, narrowed lower part of the face, short erect vertebral column, great mental acuteness, and restless vivacity, has emerged from the holes of the earth, the recesses of the forests and wastes, into which it had been swept before the more powerful blue-eyed Caucasian; and being a smaller, more easily subsisting animal, has, by starving and eating out, been gradually undermining the breed of its former conquerors. The changes which have been taking place in France, and which, in many places, leave now scarcely a trace of the fine race which existed twenty centuries ago, may, however, in part, be accounted for by the admixture of the Caucasian and Keltic tending more to the character of the latter, from the latter being a purer and more fixed variety, and nearer the original type or medium standard of man; and from the warm dry plains of France (much drier from cultivation and the reduction of the forests), having considerable influence to increase this bias. In some of the south-eastern departments, more immediately in the tide of the ingress of the Caucasian, where the purest current has latest flowed, and the climate is more suitable, and also in some of the maritime districts, where the air is moister, and to which they have been seaborn at a later period, the Caucasian character is still prominent. Something of this, yet not so general, is occurring in Britain, where the fair bright-blooded race is again giving place to the darker and more sallow. This may, however, be partly occasioned by more of artificial heat and shelter and other consequences of higher civilization. There seems to be something connected with confinement and sedentary life, with morbid action of the liver, or respiratory or transpiratory organs, which tend to this change under dry and hot, and especially confined atmosphere. Perhaps imagination is also a worker here; and the colour most regarded, as snow in cold countries, black among colliers, white among bleachers, or even the dark colour of dress, may produce its peculiar impression, and our much looked-up-to Calvinistic priesthood, from the pulpit, disseminate darkness as well as light.

Our own Kelt has indubitably improved much since, *par necessité,* he took to the mountain; but, though steadily enduring, when there is mental excitement, he has acquired a distaste to dull hopeless unceasing labour, and would fare scantily and lie hard, rather than submit to the monotonous industry of the city operative, or the toil of the agricultural drudge. Though once a fugitive, the Kelt is now, in moral courage and hardihood, equal perhaps to any other, yet he still trembles to put foot on ocean.

Notwithstanding that change of place, simply, may have impression to improve the species, yet is it more to circumstances connected with this change, to which the chief part of the improvement must be referred. In the agitation which accompanies emigration, the ablest in mind and body – the most powerful varieties of the race will be thrown into their natural position as leaders, impressing the stamp of their character on the people at large, and constituting the more reproductive part; while the feebler or more improvident varieties will generally sink under the incidental hardships. When a swarm emigrates from a prosperous hive, it also will generally consist of the more adventurous stirring spirits, who, with the right of conquerors, will appropriate the finest of the indigenae which they overrun; their choice of these being regulated by personal qualities, not by the adventitious circumstances of wealth or high birth – a regard to which certainly tends to deteriorate the species, and is one of the causes which renders the noblesse of Europe comparatively inferior to the Asiatic, or rather the Christian noblesse to the Mahometan.

It has been remarked, that our finest, most acute population, exist in the neutral ground, where the Caucasian and Keltic have mixed, but this may arise from other causes than admixture. Our healthiest and poorest country borders the Highlands, and the population enjoy more of the open air. Our eastern population, north of the natural division of Flamboroughhead, are also harder and sharper featured, and keener witted, than those southward, who may be styled our fen-bred. There is no doubt more of Keltic blood mingled with the north division; but the sea-born breeds have also been different, those more northerly being Scandinavian, and the more southerly consisting of the native of Lower Germany and the heavy Fleming. The placid-looking Englishman, more under the control of animal enjoyment, though perhaps not so readily acute, excels in the no less valuable qualities of constancy and bodily powers of exertion; and when properly taught under high division of labour, becomes a better operative in his particular employment, and even will sometimes extend scientific discovery further, than his more mercurial northern neighbour, who, from his quick wits being generally in advance of his manual practice, seldom attains to the dexterity which results from the combination of continued bodily action and restricted mental application. There exists, however, very considerable intellectual capacity in this English breed, but it too frequently is crushed under the preponderance of the animal part, affording that purest specimen of vulgarity, the English clown. But, independently of climate and breed, a great part of the low Englander's obtuseness is referable to his being entailed lord of the soil, under poor-rate law, contravening a natural law (see note B), so that, when unsuccessful or out of employment, he, without effort to obtain some new means of independent subsistence, sinks into the parish or work-house labourer. On the contrary, the Scotsman, with no resource but in himself, with famine always in the vista, as much in his view as a principle of action in material affairs as his strong perception of the right in moral, and also under the stimulus of a high pride, leaves no means untried at home; and, when fairly starved out of his native country, among various resources, often invades the territory of his more easy-minded southern neighbour, where his acuteness seldom fails to find out a convenient occupation, in which manual dexterity is second to economy and forethought – his success exciting the wonder and envy of the dull-witted native.

It would appear, that the finest portion, at least apparently so, of the north temperate zone, between the parallels of 30° and 48° latitude, when nearly of the level of the ocean, is not so favourable for human existence as the more northern part between 50° and 60°, or even the torrid zone. The native of the north of Europe has a superior development of person, and a much longer reproductory life than the native of the south, which more than counterbalances the earlier maturity of the latter in power of increase. Independent of the great current of population setting south in the northern part of the temperate zone, there seems even to be some tendency to a flux northward, from the confines of the torrid; but this arises rather from the unsteadiness of the seasons, and consequent deficit of food, at particular times, than from a steady increase of population.*

◊

[Page 381] Throughout this volume, we have felt considerable inconvenience, from the adopted dogmatical classification of plants, and have all along been floundering between species and variety, which certainly under culture soften into each other. A particular conformity, each after its own kind, when in a state of nature, termed species, no doubt exists to a considerable degree. This conformity has existed during the last forty centuries. Geologists discover a like particular conformity – fossil species – through the deep deposition of each great epoch, but they also discover an almost complete difference to exist between the species or stamp of life, of one epoch from that of every other. We are therefore led to admit, either of a repeated miraculous creation; or of a power of change, under a change of circumstances, to belong to living organized matter, or rather to the congeries of inferior life, which appears to form superior. The derangements and changes in organized existence, induced by a change of circumstance from the interference of man, affording us proof of the plastic quality of superior life, and the likelihood that circumstances have been very different in the different epochs, though steady in each, tend strongly to heighten the probability of the latter theory.

When we view the immense calcareous and bituminous formations, principally from the waters and atmosphere, and consider the oxidations and depositions which have taken place, either gradually, or during some of the great convulsions, it appears at least probable, that the liquid elements containing life have varied considerably at different times in composition and in weight; that our atmosphere has contained a much greater proportion of carbonic acid or oxygen; and our waters, aided by excess of carbonic acid, and greater heat resulting from greater density of atmosphere, have contained a greater quantity of lime and

*Omitted here are pages 376 through 380.

other mineral solutions. Is the inference then unphilosophic, that living things which are proved to have a circumstance-suiting power – a very slight change of circumstance by culture inducing a corresponding change of character – may have gradually accommodated themselves to the variations of the elements containing them, and, without new creation, have presented the diverging changeable phenomena of past and present organized existence.

The destructive liquid currents, before which the hardest mountains have been swept and comminuted into gravel, sand, and mud, which intervened between and divided these epochs, probably extending over the whole surface of the globe, and destroying nearly all living things, must have reduced existence so much, that an unoccupied field would be formed for new diverging ramifications of life, which, from the connected sexual system of vegetables, and the natural instincts of animals to herd and combine with their own kind, would fall into specific groups, these remnants, in the course of time, moulding and accommodating their being anew to the change of circumstances, and to every possible means of subsistence, and the millions of ages of regularity which appear to have followed between the epochs, probably after this accommodation was completed, affording fossil deposit of regular specific character.

There are only two probable ways of change – the above, and the still wider deviation from present occurrence, – of indestructible or molecular life (which seems to resolve itself into powers of attraction and repulsion under mathematical figure and regulation, bearing a slight systematic similitude to the great aggregations of matter), gradually uniting and developing itself into new circumstance-suited living aggregates, without the presence of any mould or germ of former aggregates, but this scarcely differs from new creation, only it forms a portion of a continued scheme or system.

In endeavouring to trace, in the former way, the principle of these changes of fashion which have taken place in the domiciles of life, the following questions occur: Do they arise from admixture of species nearly allied producing intermediate species? Are they *the diverging ramifications* of the living principle under modification of circumstance? Or have they resulted from the combined agency of both? Is there only one living principle? Does organized existence, and perhaps all material existence, consist of one Proteus principle of life capable of gradual circumstance-suited modifications and aggregations, without bound under the solvent or motion-giving principle, heat or light? There is more beauty and unity of design in this continual balancing of life to circumstance, and greater conformity to those dispositions of nature which are manifest to us, than in total destruction and new creation. It is improbable that much of this diversification is owing to commixture of species nearly allied, all change by this appears very limited, and confined within the bounds of what is called Species; the progeny of the same parents, under great difference of circumstance, might, in several generations, even become distinct species, incapable of co-reproduction.

The self-regulating adaptive disposition of organized life may, in part, be traced to the extreme fecundity of Nature, who, as before stated, has, in all the varieties of her offspring, a prolific power much beyond (in many cases a thousandfold) what is necessary to fill up the vacancies caused by senile decay. As the field of existence is limited and pre-occupied, it is only the hardier, more robust, better suited to circumstance individuals, who are able to struggle forward to maturity, these inhabiting only the situations to which they have superior adaptation and greater power of occupancy than any other kind; the weaker, less circumstance-suited, being prematurely destroyed. This principle is in constant action, it regulates the colour, the figure, the capacities, and instincts; those individuals of each species, whose colour and covering are best suited to concealment or protection from enemies, or defence from vicissitude and inclemencies of climate, whose figure is best accommodated to health, strength, and support; whose capacities and instincts can best regulate the physical energies to self-advantage according to circumstances – in such immense waste of primary and youthful life, *those* only come forward to maturity from the strict ordeal by which Nature tests their adaptation to her standard of perfection and fitness to continue their kind by reproduction.

From the unremitting operation of this law acting in concert with the tendency which the progeny have to take the more particular qualities of the parents, together with the connected sexual system in vegetables, and instinctive limitation to its own kind in animals, a considerable uniformity of figure, colour, and character, is induced, constituting species; the breed gradually acquiring the very best possible adaptation of these to its condition which it is susceptible of, and when alteration of circumstance occurs, thus changing in character to suit these as far as its nature is susceptible of change.

This circumstance-adaptive law, operating upon the slight but continued natural disposition to sport in the progeny (seedling variety), does not preclude the supposed influence which volition or sensation may have over the configuration of the body. To examine into the disposition to sport in the progeny, even when there is only one parent, as in many vegetables, and to investigate how much variation is modified by the mind or nervous sensation of the parents, or of the living thing itself during its progress to maturity; how far it depends upon external circumstance, and how far on the will, irritability and muscular exertion, is open to examination and experiment. In the first place, we ought to investigate its dependency upon the preceding links of the particular chain of life, variety being often merely types or approximations of former parentage; thence the variation of the family, as well as of the individual, must be embraced by our experiments.

This continuation of family type, not broken by casual particular aberration, is mental as well as corporeal, and is exemplified in many of the dispositions or instincts of particular races of men. These innate or continuous ideas or habits, seem proportionally greater in the insect tribes, those especially of shorter revolution; and forming an abiding memory, may resolve much of the enigma of instinct, and the foreknowledge which these tribes have of what is necessary to completing their round of life, reducing this to knowledge, or impressions, and habits, acquired by a long experience. This greater continuity of existence, or rather continuity of perceptions and impressions, in insects, is highly probable; it is even difficult in some to ascertain the particular stops when each individuality commences, under the different phases of egg, larva, pupa, or if much conscious of individuality exists. The continuation of reproduction for several generations by the females alone in some of these tribes, tends to the probability of the greater continuity of existence, and the subdivisions of life by cuttings, at any rate must stagger the advocate of individuality.

Among the millions of *specific varieties* of living things which occupy the humid portion of the surface of our planet, as far back as can be traced, there does not appear, with the exception of man, to have been any particular engrossing race, but a pretty fair balance of powers of occupancy, – or rather, most wonderful variation of circumstance parallel to the nature of every species, as if circumstance and species had grown up together. There are indeed several races which have threatened ascendency in some particular regions, but it is man alone from whom any general imminent danger to the existence of his brethren is to be dreaded.

As far back as history reaches, man had already had considerable influence, and had made encroachments upon his fellow denizens, probably occasioning the destruction of many species, and the production and continuation of a number of varieties or even species, which he found more suited to supply his wants, but which, from the infirmity of their condition – not having undergone selection by the law of nature, of which we have spoken, cannot maintain their ground without his culture and protection.

It is, however, only in the present age that man has begun to reap the fruits of his tedious education, and has proven how much "knowledge is power." He has now acquired a dominion over the material world, and a consequent power of increase, so as to render it probable that the whole surface of the earth may soon be overrun by this engrossing anomaly, to the annihilation of every wonderful and beautiful variety of animated existence, which does not administer to his wants principally as laboratories of preparation to befit cruder elemental matter for assimilation by his organs.

•

1834

Charles Lyell
Whether Species Have a Real Existence in Nature

[Charles Lyell, (Recapitulation to) "Whether Species Have a Real Existence in Nature," *Principles of Geology*. Fourth Edition. London: John Murray, 1834, Vol. 3, Book 3, Ch. 4, pp. 20-21. Except for punctuation, the six inferences are identical to those in the First Edition, 1832, Vol. 2, pp. 64-65.]

....WE MAY DRAW the following inferences in regard to the reality of *species* in nature:–

1st. That there is a capacity in all species to accommodate themselves, to a certain extent, to a change of external circumstances, this extent varying greatly, according to the species.

2dly. When the change of situation which they can endure is great, it is usually attended by some modifications of the form, colour, size, structure, or other particulars; but the mutations thus superinduced are governed by constant laws, and the capability of so varying forms part of the permanent specific character.

3dly. Some acquired peculiarities of form, structure, and instinct, are transmissible to the offspring; but these consist of such qualities and attributes only as are intimately related to the natural wants and propensities of the species.

4thly. The entire variation from the original type, which any given kind of change can produce, may usually be effected in a brief period of time, after which no farther deviation can be obtained by continuing to alter the circumstances, though ever so gradually; indefinite divergence, either in the way of improvement or deterioration, being prevented, and the least possible excess beyond the defined limits being fatal to the existence of the individual.

5thly. The intermixture of distinct species is guarded against by the aversion of the individuals composing them to sexual union, or by the sterility of the mule offspring. It does not appear that true hybrid races have ever been perpetuated for several generations, even by the assistance of man; for the cases usually cited relate to the crossing of mules with individuals of pure species, and not to the intermixture of hybrid with hybrid.

6thly. From the above considerations, it appears that species have a real existence in nature; and that each was endowed, at the time of its creation, with the attributes and organization by which it is now distinguished.

•

1835

Edward Blyth
An Attempt to Classify Varieties of Animals

[Edward Blyth, "An Attempt to Classify the 'Varieties' of Animals, with Observations on the Marked Seasonal and Other Changes Which Naturally Take Place in Various British Species, and Which Do Not Constitute Varieties," in *The Magazine of Natural History, and Journal of Zoology, Botany, Mineralogy, Geology, and Meteorology*, 8 (1835), pp. 40-53]

THE APPELLATION "variety" being very commonly misapplied to individuals of a species, which are merely undergoing a regular natural change, either progressing from youth to maturity, or gradually shifting, according to fixed laws, their colours with the seasons, I conceive that it will be useful to some, to point out a few of the less generally known changes which naturally take place in various British animals; some few of which appear to have been hitherto overlooked, and others to have been described incorrectly.

The term "variety" is understood to signify a departure from the acknowledged type of a species, either in structure, in size, or in colour; but is vague in the degree of being alike used to denote the slightest individual variation, and the most dissimilar breeds which have originated from one common stock. The term is, however, quite inapplicable to an animal in any state of periodical change natural to the species to which it belongs.

Varieties require some classification; and though I feel myself hardly adequate to the task, I shall here propose to arrange them under four principal heads; in the hope that this endeavour will induce some naturalists, more competent than myself, to follow out this intricate and complicated subject, into all its details.

I would distinguish, then, what are called varieties, into *simple variations, acquired variations, breeds,* and *true varieties.* These appear, in general, sufficiently distinct, although the exact limits of each are sometimes very difficult to be assigned. Indeed, in many cases they only differ in degree, and in others they may be all combined in one individual. Moreover, the varieties of either class, have a much greater tendency to produce varieties of another class, than the typical animals of a species have to produce any sort of variety.

I. *Simple Variations.* – The first class, which I propose to style *simple or slight individual variations,* differs only in degree from the last, or *true varieties*; and consists of mere differences of colour or of stature, unaccompanied by any remarkable structural deviation; also of slight individual peculiarities of any kind, which are more or less observable in all animals, whether wild or tame, and which, having a tendency to perpetuate themselves by generation, may, under particular circumstances, become the origin of true *breeds* (which constitute my third class of varieties), but which, in a state of nature, are generally lost in the course of two or three generations. Albinoes belong to this first division, and also the other numerous anomalies mentioned in VII. 589-591. 593-598. These *simple variations* occur both in wild and in domestic animals, but are much more frequent in the latter, and are commonly observed in all *breeds* and *true varieties.*

Among the Mammalia, total or partial absence of colour is always, I believe, continued through life; excepting, of course, the cases of mere seasonal change; and, in this class of animals generally, perfect albinoes are much more numerous than among birds. Perfect albinoes are peculiar to warm-blooded animals, and in them there is a total deficiency of colouring matter in the rete mucosum, and, consequently, in the fur, and even the pigmentum nigrum of the eye is entirely wanting. In birds, these *perfect* albinoes are extremely rare, although several instances have been recorded in VII.593.–598. There are three sorts, however, of true permanent albinoes, which may be thus designated:–1. *Perfect Albinoes*; which are entirely white, and in which the eyes appear crimson, from the total want of colouring matter, rendering the minute bloodvessels visible: 2. *Semi-Albinoes*; which are either white or of a pale colour all over, and in which the irises are always paler than usual, and not unfrequently blue [I.66.178.]: and, 3. *Partial Albinoes*; which are partly of the natural colour, but are more or less mottled *permanently* with white; and in which, if a white patch surrounds the eye, the pigmentum of that organ is commonly wanting.

I have thus observed a rabbit, one eye of which was red, and the other dark hazel; but such instances are of very rare occurrence, although (and it is a curious fact) rabbits are often seen wholly white, with the exception of a small patch around each eye; which organ, consequently, is of the usual dark colour. Albinoes, when paired together, as is well known, produce chiefly albino offspring, and a *breed* of them may thus be perpetuated; but, even in a domestic state, they not unfrequently produce young of the usual colour; and, if paired with an ordinary individual, they sometimes produce partial albinoes, or semi-albinoes [I. 178.], and occasionally, if the original colour be brown (as in the case of mice or rabbits*), a black, sandy, or slate-coloured offspring, or an individual with one of these colours more or less varied with white, is produced; but, in the majority of instances, the young wholly resemble one of their parents, and the *preponderance* is decidedly in favour of the natural hue. The coloured offspring of an albino, however, even if matched with another coloured individual, has still a tendency to produce albinoes†, and this fact has been noticed in the human species; but, as Mr. Lawrence observes on the subject (in his *Lectures on the Physiology, Zoology, and the Natural History of Man*), "the disposition to change is 'generally' exhausted in one individual, and the characters of the original stock return, unless the variety is kept up by the precaution above mentioned, of excluding from the breed all which have not the new characters. Thus, when African albinoes intermix with the common race, the offspring generally is black," &c. These observations apply alike to all *simple* or individual *variations,* and to most other varieties, and afford one of many reasons why marked breeds are in a state of nature so rarely perpetuated. There is yet, however, before quitting this subject, another sort of albino to be considered, which, I believe, is peculiar to the feathered race, and which is not, like the others, permanent; these, therefore, I shall denominate *temporary* albinoes. Most of the pale, white, and pied varieties of birds, which are produced in a state of nature, are of this kind. A friend informs me that a perfectly white lark in his possession moulted, and became of the ordinary hue. I lately shot a sparrow which was all over of a very pale brown, or cream colour; it was moulting, and some of the new feathers that were coming were of the usual colour, and others

*These observations are chiefly deduced from the results of some experiments with mice and rabbits.

†Of seven young rabbits thus produced, two were albinoes, one black, and the remainder of the usual colour.

were of a pure white: on the next moult, probably, no more white feathers would have appeared. Of a brood of young robins which frequented my garden, two were white, one partially so, and one of the usual mottled brown; these all moulted into the ordinary colour. I could add other instances to the list, especially amongst domestic poultry. But it does not hence follow that among wild birds there are *no* permanently white or pied varieties; or, in other words, no true partial and semi-albinoes. A blackbird with a white head has now inhabited a garden in this neighbourhood for three successive years; and if the cupidity of collectors did not mark out every white or pied bird for destruction, I doubt not that I should have been able to have furnished some other similar instances of *permanent* variation.

II. *Acquired Variations.* – The second class of varieties which I would designate thus, comprises the various changes which, in a single individual, or in the course of generations, are *gradually* brought about by the operation of known causes: such as the greater or less supply of *nutriment*; the influence of particular *sorts* of food; or, either of these combined with the various privations consequent upon *confinement*; which changes would as gradually and certainly disappear if these causes were removed.

Redundance or deficiency of nutriment affects chiefly the stature of animals. Those herbivorous quadrupeds which browse the scanty vegetation on mountains are invariably much smaller than their brethren which crop the luxuriant produce of the plains; and although the cattle usually kept in these different situations are of diverse breeds; yet either of the breeds gradually removed to the other's pasture, would, in two or three generations, acquire many of the characters of the other, would increase or degenerate in size, according to the supply of nutritious food; though, in either case, they would most probably soon give birth to *true varieties* adapted to the change. In this instance, *temperature* appears only to exert a secondary influence. The Iceland breed of sheep, which feeds on the nutritious lichens of that island, is of large size; and, like the other ruminant animals which subsist on similar food, is remarkable for an extraordinary development of horns. Another example of *acquired variation,* dependent solely on the supply of nutriment, may be observed in the deciduous horns of the deer family, which are well known to be large or small according to the quality of their food. That *temperature* also does exert an influence greater or less, according to the species of animal, is very evidently shown in the case of the donkey*, of which there are no breeds, nor true varieties, and but very few simple variations [VII. 590.]: this animal is every where found large or small, according to the *climate* it inhabits.

The influence of particular *sorts* of food may be exemplified by the well-known property of madder (Rùbia tinctòrum), which colours the secretions, and tinges even the bones of the animals which feed on it of a blood-red colour; and, as another familiar instance, may be cited the fact, equally well known, of bullfinches, and one or two other small birds, becoming wholly black when fed entirely on hempseed. I have known, however, this change to take place in a bird (the small aberdevatt finch, so common in the shops), which had been wholly fed on canary seed; yet this by no means invalidates the fact, so often observed, of its being very frequently brought about by the direct influence of the former diet. In several instances which have fallen under my own observation, feeding only on hempseed has invariably superinduced the change.†

The most remarkable of acquired variations are those brought about in animals in a state of confinement or domestication: in which case an animal is supplied regularly with abundance of very nutritious, though often unnatural, food, without the trouble and exertion of having to seek for it, and it becomes, in consequence, bulky and lazy, and in a few generations often very large; while the muscles of the organs of locomotion, from being but little called into action, become rigid and comparatively powerless, or are not developed to their full size. The common domestic breeds of the rabbit, ferret, guinea-pig, turkey, goose, and duck, are thus probably only acquired variations, which, from the causes above-mentioned, have, in the course of generations, become much larger and heavier (excepting, however, in the case of the turkey) than their wild prototypes, and less fitted for locomotion; but which, if turned loose into their natural haunts, would most probably return, in a very few generations, to the form, size, and degree of

---

*For some curious remarks on this subject, see the excellent article "Ass" in Partington's *Cyclopaedia of Natural History*.

†I have not heard, however, that wild bullfinches, hawfinches, and other birds liable to be thus affected, are more commonly found black in localities where hemp is much grown. Amongst others, the skylark and woodlark are very susceptible of the influence of this food.

locomotive ability proper to the species when naturally conditioned.* The crested varieties of domestic geese and ducks, and the hook-billed variety of the latter, are, however, in all probability, *true varieties*; and what are called "lob-eared" rabbits may be either a *true variety*, or a *breed*. The various slight diversities, which I call *simple variations*, are very common in the present class of varieties; and there is also in them a great tendency to produce what I call *true varieties*, as well as those slighter deviations, which, by particular management, may be increased into the sort of variety I denominate *breeds*.

III. *Breeds* are my third class of varieties; and though these may possibly be sometimes formed by accidental isolation in a state of nature, yet they are, for the most part, artificially brought about by the *direct* agency of *man*. It is a general law of nature for all creatures to propagate the like of themselves: and this extends even to the most trivial minutiae, to the slightest individual peculiarities; and thus, among ourselves, we see a family likeness transmitted from generation to generation. When two animals are matched together, each remarkable for a certain given peculiarity, no matter how trivial, there is also a decided tendency in nature for that peculiarity to *increase*; and if the produce of these animals be set apart, and only those in which the same peculiarity is most apparent, be selected to breed from, the next generation will possess it in a still *more* remarkable degree; and so on, till at length the variety I designate a *breed*, is formed, which may be very unlike the original type.

---

*[*A Tame Duck which flies with the same Power, and at the same Height, as a Crow.* (H. S., in I. 378.) – Was not this duck a wild one? I am led to ask this question from having myself witnessed a similar instance. I had often seen a duck, which I had taken to be a tame one, flying about, and always returning to the farm to which it belonged. On enquiry, I found that this duck had been taken, when a duckling, from the nest of a wild duck, and began to fly as soon as it was full grown. The case which H. S. mentions might probably be accounted for in the same manner, as it is by no means likely that so unwieldy a bird as the tame duck should think of trying its wings, after its ancestors had for so many successive generations been satisfied with walking and swimming, and fly "with the same power, and at the same height, as a crow." –*W.H.H. Postmark, Burton on Trent, Oct. 8, 1834.*

The late Rev. Lansdown Guilding had remarked as follows on the case stated by H.S.:

Domestic birds, from flying little, have their muscles relaxed, or, perhaps, they never acquire their natural strength, for want of exercise. I have observed the geese in Worcestershire, in harvest time, to take very long flights; but, though they went on boldly, they never ascended very far into the air. – *Lansdown Guilding. St. Vincent, May 1, 1830.*]

The examples of this class of varieties must be too obvious to need specification: many of the varieties of cattle, and, in all probability, the greater number of those of domestic pigeons, have been generally brought about in this manner. It is worthy of remark, however, that the original and typical form of an animal is in great measure kept up by the same identical means by which a true *breed* is produced. The original form of a species is *unquestionably* better adapted to its *natural* habits than any modification of that form; and, as the sexual passions excite to rivalry and conflict, and the stronger must always prevail over the weaker, the latter, in a state of nature, is allowed but few opportunities of continuing its race. In a large herd of cattle, the strongest bull drives from him all the younger and weaker individuals of his own sex, and remains sole master of the herd; so that all the young which are produced must have had their origin from one which possessed the maximum of power and physical strength; and which, consequently, in the struggle for existence, was the best able to maintain his ground, and defend himself from every enemy. In like manner, among animals which procure their food by means of their agility, strength, or delicacy of sense, the one best organised must always obtain the greatest quantity; and must, therefore, become physically the strongest, and be thus enabled, by routing its opponents, to transmit its superior qualities to a greater number of offspring. The same law, therefore, which was intended by Providence to keep up the typical qualities of a species, can be easily converted by man into a means of raising different varieties; but it is also clear that, if man did not keep up these breeds by regulating the sexual intercourse, they would all naturally soon revert to the original type. Farther, it is only on this principle that we can satisfactorily account for the degenerating effects said to be produced by the much-censured practice of "breeding in and in."* There would almost seem, in some species, to be a tendency, in every separate family, to some particular kind of deviation; which is only counteracted by the various crossings which, in a state of nature, must take place, and by the above-mentioned law, which causes each race to be chiefly propagated by the most typical and perfect individuals.

---

*See, however, a good practical article on this subject, entitled "Breeding," in one of the forthcoming numbers of the now publishing edition of Miller's *Dictionary of Gardening and Rural Economy.*

IV. *True Varieties.* The last of these divisions to which I more peculiarly restrict the term *variety,* consists of what are, in fact, a kind of deformities, or monstrous births, the peculiarities of which, from reasons already mentioned, would very rarely, if ever, be perpetuated in a state of nature; but which, by man's agency, often become the origin of a new race. Such, for example, is the breed of sheep, now common in North America, and known by the name of *Ancons**, or *otter* sheep. A ewe produced a male lamb of peculiar form, with a long body, and short and crooked limbs: the offspring of this animal, with ordinary females, was found sometimes to resemble the one parent, and sometimes the other; but did not usually blend the characters of each; and, in the cases of twins, the two lambs were often equally diverse with their parents. This variety was extensively propagated, in consequence of being less able to jump over fences than the ordinary breeds of sheep. The solidungular ["donkey-footed"] variety of swine, tailless cats, back-feathered, five-toed, and rumpless fowls, together with many sorts of dogs, and probably, also, the race of fan-tailed pigeons, are other striking examples of *true varieties.*

The deviations of this kind do not appear to have any tendency to revert to the original form: this, most probably, could only be restored, in a direct manner, by the way in which the variety was first produced.

To this class may be also referred, with more than probability, some of the more remarkable varieties of the human species. With regard to colour, we know that temperature exerts no *permanent* gradual influence whatever: white races remain unchanged at slight elevations within the tropics; and the natives of Boothia Felix are very dark; the swarthy inhabitants of Mauritania are a white race, and their sunburnt hue is merely an *acquired variation,* which is not transmissible by generation, and which does not extend to those parts which are not exposed to the sun. The colouring principle of black races is inherent in them, and is quite independent of external agency; is even darkest in some parts which are the least exposed, and *vice versâ.* The Ethiopian race is nowhere more black than in the vicinity of the Cape of Good Hope, where the crops are sometimes injured by the winter's frost. Strangely enough, this invariableness of colour constitutes about, perhaps, the most fixed character of these races.

There is one fact, however, here to be observed, which is very well worthy of attention; and this is, that coloured varieties appear to have

---

**Agkōn,* an elbow, from the crooked form of the forelegs. See Lawrence's *Lectures,* p. 447, 448.

been chiefly produced in hot countries; which seems almost to induce the conclusion that they were originally efforts of nature, to enable the skin to withstand the scorching produced by exposure to the burning rays of a tropical sun.* How far the structural peculiarities of the negro and other races may not, in some cases, be the effects of *breed,* it would be impossible, perhaps, now to ascertain, and would be worse than presumption, in a novice like myself, to try to determine. Wherever a black individual was produced, especially among rude nations, if the breed was continued at all, the natural aversion it would certainly inspire would soon cause it to become isolated, and, before long, would, most probably, compel the race to seek for refuge in emigration. That no example, however, of the first production of a black variety has been recorded, may be ascribed to various causes; it may have only taken place once since the creation of the human race, and that once in a horde of tropical barbarians remote from the then centres of comparative civilisation, where no sort of record would have been preserved. But it is highly probable that analogous-born varieties may have given rise to the Mongolian, Malay, and certain others of the more diverse races of mankind; nay, we may even suppose that, in some cases, the difference, in the first instance, was much *greater,* and was considerably modified by the intermixture which must have taken place in the first generations. The mixed offspring of two different varieties of man thus generally blends the characters of each; though instances are not wanting of its *entirely* resembling (like the mixed produce of an ancon sheep) either one or the other of its parents; but in this case (as in the albino) the perfect characters of the other parent frequently show themselves in the next generation. I am entering, however, into a wide field, already well trodden by many philosophers; and the subject is already probably pretty well understood by the great majority of readers. Those who are not so familiar with it, will find it ably treated in various works; especially in Dr. Pritchard's work on man, and in the published *Lectures on the Natural History of Man,* by Lawrence: some sound and excellent remarks on *varieties* will also be found in the second volume of Lyell's *Principles of Geology.*

Still, however, it may not be impertinent to remark here, that, as in the brute creation, by a wise provision, the typical characters of a

---

*See Dr. Stark "on the influence of colour on heat and odours," in Jameson's *Edinburgh Journal* for July, 1834; also Professor Powell's reply to it, in the number for October, 184. [1834?]

species are, in a state of nature, preserved by those individuals chiefly propagating, whose organisation is the most perfect, and which, consequently, by their superior energy and physical powers, are enabled to vanquish and drive away the weak and sickly, so in the human race degeneration is, in great measure, prevented by the innate and natural preference which is always given to the most comely; and this is the principal and main reason why the varieties which are produced in savage tribes, must generally either become extinct in the first generation, or, if propagated, would most likely be left to themselves, and so become the origin of a new race; and in this we see an adequate cause for the obscurity in which the origin of different races is involved. In a civilised state of society there are other inducements, besides personal attractions, and a new variety in this case, unless very *outré* indeed, would be gradually merged, and in a few generations would disappear entirely by intermixture with the common race. The inferior animals appear not to have the slightest predilection for superior personal appearance; the most powerful alone becomes the favourite. Instances of this are not rare in the breeds of dogs.

The above is confessedly a hasty and imperfect sketch, a mere approximation towards an apt classification of "varieties;" but if it chance to meet the eye, and be fortunate enough to engage the attention, of any experienced naturalist, who shall think it worth his while to follow up the subject, and produce a better arrangement of these diversities, my object in indicting the present article will be amply recompensed.

Here, however, I may observe, that the classification I have proposed for specific deviations in the animal creation, is equally applicable to those of the vegetable. The "varieties" in both are strictly analogous.

I come now to the second division of my subject, which is to point out

Some Periodical and other Changes of Appearance, which naturally take place in various British Animals, and which do not constitute Varieties. —Among our native Mammàlia, I know of three principal modes whereby a change of colour is brought about; namely, an actual shedding of the coat; a partial shedding of the coat; and an actual change of colour in the coat itself.

I. As an example of change of appearance produced by actual shedding of the coat, may be instanced the fallow deer (Cèrvus Dàma), whose white spots disappear with the annual casting of its coat in autumn.

II. Partial shedding of the coat takes place in those animals which acquire in autumn a covering of two different kinds: one long, downy, and warm, which is shed in spring; the other short and glossy, which is retained. This change of appearance is exemplified in the common water shrew (Sòrex fòdiens), the short summer coat of which is much blacker than the longer downy covering which conceals this in winter. In this little animal the additional winter coat is shed about the latter end of March, or beginning of April; and does not take place uniformly, but progressively, beginning on the head, and ceasing at the hinder extremities; and exhibiting in its progress, throughout, a well-defined line of separation. Animals which (as the British Mustelìnae) have two sorts of fur, the *shorter* of which is the more warm and downy, do not undergo this change, but retain both sorts throughout the year. In these the young have only one kind, which is close and woolly; as is well exemplified in the common polecat (Putòrius Fùro), the young of which are of a very uniform dark brown, and very unlike the old animals.

III. Actual change of colour in the coat itself is exhibited in the appearance of the fallow deer's white spots in spring, and in the case of the mountain hare (Lèpus variábilis), which is in summer grey, adapted to the hue of the lichens on which it squats; and in winter white, hardly to be discerned upon the snow. The same change also takes place in the stoat or ermine (Putòrius ermìneus), although this is doubted by Mr. Berry (VII. 591.).* In mild winters, such as we have of late experienced in the South of England, but few of the stoats become white, and some of these not until the latter part of the season. The change takes place quickly, but not uniformly, the animal assuming for a short time a pied appearance; but I have not succeeded in ascertaining whether it is accelerated by sudden cold, as the animals are not always to be seen exactly when we want them. One perfectly changed, however, was seen in this neighbourhood soon after the one or two days of very cold weather in the beginning of last October. In reference to Mr. Berry's communication, I may observe, that in many dozens of stoats which I

*This gentleman should have mentioned, in his account of the white stoats seen in summer, whether the tail was white or black. If the former, they were doubtless albinoes; if the latter, some constitutional debility may have prevented them from resuming their natural hues. I have seen white stoats late in March, but never after this. Both in these and in the white ferret (a domestic albino variation of the polecat) a decided tinge of yellow is always more or less noticeable.

have seen in summer, I have never yet seen a white one; whereas in winter, I have seen in the same neighbourhoods a considerable number of white stoats. Where the climate is more excessive, and the transitions of the seasons are more sudden, this change is much more likely to take place generally. In the fur countries, the ermine's change of hue is, I believe, most regular.

There has been, strangely enough, a difference of opinion among naturalists, as to whether these seasonal changes of colour were intended by Providence as an adaptation to change of temperature*, or as a means of preserving the various species from the observation of their foes, by adapting their hues to the colour of the surface; against which latter opinion it has been plausibly enough argued, that "nature provides for the preyer as well as for the prey." The fact is, they answer *both* purposes; and they are among those striking instances of *design,* which so clearly and forcibly attest the existence of an omniscient great First Cause. Experiment demonstrates the soundness of the first opinion; and sufficient proof can be adduced to show that the other is also sound. Some arctic species are white, which have no enemy to fear, as the polar bear, the gyrfalcon, the arctic eagle-owl, the snowy owl, and even the stoat; and therefore, in these, the whiteness can only be to preserve the temperature of their bodies [VI. 79.]; but when we perceive that the colour of nocturnal animals, and of those defenceless species whose habits lead them to be much exposed, especially to enemies from above, are *invariably* of the same colour with their respective natural haunts, we can only presume that this is because they should not appear too conspicuous to their enemies. Thus, in the eloquent language of Mr. Mudie†, who, however, advocates the first opinion, "the ptarmigan is lichen rock in summer, hoar frost in autumn, and snow in winter. Grouse are brown heather, black game are peat bank and shingle, and partridges are clods and withered stalks, all the year round." So, also, on the Continent, the common red-legged partridge (Erýthropus vulgàris) is of the colour of the gravelly and sandy soils on which it is found. So, also, are the different larks, the common quail, the various snipes, and all the other ground squatters, of the hue of their peculiar localities. So, also, are the numerous small Grallatòres which haunt the margin of the ocean, adapted to the colour of the sand. So, also, are those sylvan birds, which quit the dense

---

*See Dr. Stark's paper, before cited, in Jameson's *Philosophical Journal* for July, 1834. [See *M.N.H.*, vi.79.]

†See Mudie's *Feathered Tribes of the British Islands,* i. 50.

umbrage of healthy growing trees, to seek their food and expose themselves on bare trunks and leafless decaying branches, of the hue of their particular haunts. "So exquisitely are they fitted for their office," says Mr. Mudie*, "that the several woodpeckers vary in tint with the general colours of the trees which they select. If it is an alternation of green moss, yellow lichen, and ruby tinted cups, with here and there a spot of black, then the green woodpecker comes in charge; but if it is the black and white lichens of the alpine forest or the harsh-juiced tree, then we may look for the spotted races upon the bark." The wryneck is the colour of the lichened branch; and the night swallow and the owls resemble their peculiar places of concealment. So, also, the gayer colours of nocturnal moths are always on the hinder wings†, and the anterior, which, when they rest, conceal these, are adapted to the hues of the various places where by day they are found: even the bright upper wings of the tiger moths (Arctia Càja, and A. vīllica) are with difficulty recongised upon a lichened bank or paling.‡ It is curious, indeed, the resemblance which subsists between the colours of nocturnal birds and night Lepidóptera; the buff tip moth (Pygaera bucéphala) thus reminds us of the barn owl (Stríx vulgàris); and the goat moth (Cóssus Lignipérda), and a host of others, are similar in their tints to most of the Strígidae: in both cases they are doubtless intended for the same purpose, that of concealment. It would indeed be easy to extend this list of examples considerably further; but I shall only now mention the common hare, which, when in form, would hardly ever be seen were it not for its brilliant eye; if its eye were closed, which it probably was before its quick sense of hearing had warned it of our approach, it would almost always, perhaps, wholly escape our observation. This ever continued watchfulness must have given rise to the supposition, that the hare always sleeps with its eyes open.

Seeing, therefore, so many most striking adaptations of colour to haunt, in cases where the concealment thus afforded can be the *only* purpose, I think it is not too much to infer, that the changes of colour in many arctic animals were intended by Providence for the double purpose of preserving their bodily heat, and of enabling them to elude

---

*See Mudie's *Feathered Tribes of the British Islands*, i. 190.

†Among day-flying Lepidóptera, the more gaudy colours are usually on the *fore* wings.

‡ Animals of bright and gaudy colours are generally very retiring in their habits: even the common robin mostly turns away his breast as you approach.

the observation of their enemies. Certain it is, that their *conspicuousness* would otherwise expose them to inevitable destruction. If I had here space, I could satisfactorily prove that the high-flying Falcónidae can, in most cases, only perceive their prey when it is *moving*; just, as on the sea-shore, *we* can only distinguish sanderlings when they move. Small Mammàlia which frequent open situations are rarely much abroad, except in the twilight; and ground-feeding birds are ever on the watch, and even the smaller kinds (as I have repeatedly observed) can perceive a hovering falcon *long* before it comes within the sphere of human vision; and they instantly flee to shelter, or they crouch, and lying motionless, so exactly resemble a portion of the surface, that even a hawk's eye cannot distinguish them. Why should the falcon race be endowed with such wonderful powers of enduring hunger and fatigue, if, as is said, at the elevations at which they soar, they can clearly distinguish every living object scattered over the wide expanse beneath them? It is only on such animals as are *off their guard* that they descend; or otherwise, food being so abundant, they would soon multiply to the extirpation of their prey; which, of course, would be very speedily followed by that of the preyer.

How beautifully do we thus perceive, as in a thousand other instances, the balance of nature preserved: and even here we see another reason why sickly or degenerate animals (those, I mean, which are less able to maintain the necessary vigilance) must soon disappear; and why the slightest deviation from the natural hue must generally prove fatal to the animal. How different, thus, are even *simple variations* from the seasonal changes of colour which naturally take place! Properly followed up, this subject might lead to some highly interesting and important results. It certainly points to the conclusion, that every, even the slightest, tint and marking has some decided use, and is intimately connected with the habits and welfare of the animal; and it also furnishes a satisfactory reason, why closely allied animals (or, in other words, animals of very similar form and habits) should so very commonly nearly resemble each other in their colours and in the general character of their markings.

•

1844

Robert Chambers
Vestiges of the Natural History of Creation

[Robert Chambers, *Vestiges of the Natural History of Creation.* London: John Churchill, 1844, pp. 152-154, 156-158, 164-165, 175-183, 184-187, 202-205, 219-223, 230-231]

[152-154] A CANDID CONSIDERATION of all these circumstances can scarcely fail to introduce into our minds a somewhat different idea of organic creation from what has hitherto been generally entertained. That God created animated beings, as well as the terraqueous theatre of their being, is a fact so powerfully evidenced, and so universally received, that I at once take it for granted. But in the particulars of this so highly supported idea, we surely here see cause for some re-consideration. It may now be inquired, – In what way was the creation of animated beings effected? The ordinary notion may, I think, be not unjustly described as this, – that the Almighty author produced the progenitors of all existing species by some sort of personal or immediate exertion. But how does this notion comport with what we have seen of the gradual advance of species, from the humblest to the highest? How can we suppose an immediate exertion of this creative power at one time to produce zoophytes, another time to add a few marine mollusks, another to bring in one or two conchifers, again to produce crustaceous fishes, again perfect fishes, and so on to the end? This would surely be to take a very mean view of the Creative Power – to, in short, anthropomorphize it, or reduce it to some such character as that borne by the ordinary proceedings of mankind. And yet this would be unavoidable; for that the organic creation was thus progressive through a long space of time, rests on evidence which nothing can overturn or gainsay. Some other idea must then be come to

with regard to *the mode* in which the Divine Author proceeded in the organic creation. Let us seek in the history of the earth's formation for a new suggestion on this point. We have seen powerful evidence, that the construction of this globe and its associates, and inferentially that of all the other globes of space, was the result, not of any immediate or personal exertion on the part of the Deity, but of natural laws which are expressions of his will. What is to hinder our supposing that the organic creation is also a result of natural laws, which are in like manner an expression of his will? More than this, the fact of the cosmical arrangements being an effect of natural law, is a powerful argument for the organic arrangements being so likewise, for how can we suppose that the august Being who brought all these countless worlds into form by the simple establishment of a natural principle flowing from his mind, was to interfere personally and specially on every occasion when a new shell-fish or reptile was to be ushered into existence on *one* of these worlds? Surely this idea is too ridiculous to be for a moment entertained. ....

[156-158] To a reasonable mind the Divine attributes must appear, not diminished or reduced in any way, by supposing a creation by law, but infinitely exalted. It is the narrowest of all views of the Deity, and characteristic of a humble class of intellects, to suppose him acting constantly in particular ways for particular occasions. It, for one thing, greatly detracts from his foresight, the most undeniable of all the attributes of Omnipotence. It lowers him towards the level of our own humble intellects. Much more worthy of him it surely is, to suppose that all things have been commissioned by him from the first, though neither is he absent from a particle of the current of natural affairs in one sense, seeing that the whole system is continually supported by his providence. Even in human affairs, if I may be allowed to adopt a familiar illustration, there is a constant progress from specific action for particular occasions, to arrangements which, once established, shall continue to answer for a great multitude of occasions. Such plans the enlightened readily form for themselves, and conceive as being adopted by all who have to attend to a multitude of affairs, while the ignorant suppose every act of the greatest public functionary to be the result of some special consideration and care on his part alone. Are we to suppose the Deity adopting plans which harmonize only with the modes of procedure of the less enlightened of our race? Those who would object to the hypothesis of a creation by the intervention of law,

do not perhaps consider how powerful an argument in favour of the existence of God is lost by rejecting this doctrine. When all is seen to be the result of law, the idea of an Almighty Author becomes irresistible, for the creation of a law for an endless series of phenomena – an act of intelligence above all else that we can conceive – could have no other imaginable source, and tells, moreover, as powerfully for a sustaining as for an originating power. .... *

[164-165] Assuming that organic beings are thus spread over all space, the idea of their having all come into existence by the operation of laws everywhere applicable, is only conformable to that principle, acknowledged to be so generally visible in the affairs of Providence, to have all done by the employment of the smallest possible amount of means. Thus, as one set of laws produced all orbs and their motions and geognostic arrangements, so one set of laws overspread them all with life. The whole productive or creative arrangements are therefore in perfect unity.

The general likelihood of an organic creation by law having been shewn, we are next to inquire if science has any facts tending to bring the assumption more nearly home to nature. Such facts there certainly are; but it cannot be surprising that they are comparatively few and scattered, when we consider that the inquiry is into one of nature's profoundest mysteries, and one which has hitherto engaged no direct attention in almost any quarter. ....

[175-183] These, it will be owned, are curious and not irrelevant facts; but it will be asked what actual experience says respecting the origination of life. Are there, it will be said, any authentic instances of either plants or animals, of however humble and simple a kind, having come into existence otherwise than in the ordinary way of generation, since the time of which geology forms the record? It may be answered, that the negative of this question could not be by any means formidable to the doctrine of law-creation, seeing that the conditions necessary for the operation of the supposed life-creating laws may not

* (Publisher's Note) On this ages-old "proof" of the existence of God, we feel compelled to quote Darwin himself:– "I cannot pretend to throw the least light on such abstruse problems. The mystery of the beginning of all things is insoluble by us; and I for one must be content to remain an Agnostic." (Page 94 of *The Autobiography of Charles Darwin*, ed. by Nora Barlow. New York: Harcourt, Brace and Co., 1958.) The continuing revelations today of interstellar chemistry and molecular biology promise to solve even this seemingly impenetrable mystery of the Beginning. Laplace would have been pleased.

have existed within record to any great extent. On the other hand, as we see the physical laws of early times still acting with more or less force, it might not be unreasonable to expect that we should still see some remnants, or partial and occasional workings of the life-creating energy amidst a system of things generally stable and at rest. Are there, then, any such remnants to be traced in our own day, or during man's existence upon earth? If there be, it clearly would form a strong evidence in favour of the doctrine, as what now takes place upon a confined scale and in a comparatively casual manner may have formerly taken place on a great scale, and as the proper and eternity-destined means of supplying a vacant globe with suitable tenants. It will at the same time be observed that, the earth being now supplied with both kinds of tenants in great abundance, we only could expect to find the life-originating power at work in some very special and extraordinary circumstances, and probably only in the inferior and obscurer departments of the vegetable and animal kingdoms.

Perhaps, if the question were asked of ten men of approved reputation in science, nine out of the number would answer in the negative. This is because, in a great number of instances where the superficial observers of former times assumed a non-generative origin for life, (as in the celebrated case in Virgil's fourth Georgic,) either the direct contrary has been ascertained, or exhaustive experiments have left no alternative from the conclusion that ordinary generation did take place, albeit in a manner which escapes observation. Finding that an erroneous assumption has been formed in many cases, modern inquirers have not hesitated to assume that there can be no case in which generation is not concerned; an assumption not only unwarranted by, but directly opposed to, the principles of philosophical investigation. Yet this is truly the point at which the question now rests in the scientific world.

I have no wish here to enter largely into a subject so wide and so full of difficulties; but I may remark, that the explanations usually suggested where life takes its rise without apparent generative means, always appear to me to partake much of the fallacy of the *petitio principii.* When, for instance, lime is laid down upon a piece of waste moss ground, and a crop of white clover for which no seeds were sown is the consequence, the explanation that the seeds have been dormant there for an unknown time, and were stimulated into germination when the lime produced the appropriate circumstances, appears extremely unsatisfactory, especially when we know that (as in an authentic case under my notice) the spot is many miles from where clover is cultivated, and that there is nothing for six feet below but pure peat moss, clover seeds being, moreover, known to be too heavy to be transported, as many other seeds are, by the winds. Mushrooms, we know, can be propagated by their seed; but another mode of raising them, well known to the gardener, is to mix cow and horse dung together, and thus form a bed in which they are expected to grow without any seed being planted. It is assumed that the seeds are carried by the atmosphere, unperceived by us, and, finding here an appropriate field for germination, germinate accordingly; but this is only assumption, and though designed to be on the side of a severe philosophy, in reality makes a pretty large demand on credulity. There are several persons eminent in science who profess at least to find great difficulties in accepting the doctrine of invariable generation. One of these, in the work noted below,* has stated several considerations arising from analogical reasoning, which appear to him to throw the balance of evidence in favour of the aboriginal production of infusoria,† the vegetation called mould, and the like. One seems to be of great force; namely, that the animalcules, which are supposed (altogether hypothetically) to be produced by ova, are afterwards found increasing their numbers, not by that mode at all, but by division of their bodies. If it be the nature of these creatures to propagate in this splitting or fissiparous manner, how could they be communicated to a vegetable infusion? Another fact of very high importance is presented in the following terms:– "The nature of the animalcule, or vegetable production, bears a constant relation to the state of the infusion, so that, in similar circumstances, the same are always produced without this being influenced by the atmosphere. There seems to be a certain *progressive advance in the productive powers of the infusion*, for at the first the animalcules are only of the smaller kinds, or monades, and afterwards *they become gradually larger and more complicated in their structure; after a time, the production ceases, although the materials are by no means exhausted.* When the quantity of water is very small, and the organic matter abundant, the production is usually of a vegetable nature; when there is much water, animalcules are more frequently

---

* Dr. Allen Thomson, in the article *Generation*, in Todd's Cyclopaedia of Anatomy and Physiology.

† The term aboriginal is here suggested, as more correct than spontaneous, the one hitherto generally used.

produced." It has been shewn by the opponents of this theory, that when a vegetable infusion is debarred from the contact of the atmosphere, by being closely sealed up or covered with a layer of oil, no animalcules are produced; but it has been said, on the other hand, that the exclusion of the air may prevent some simple condition necessary for the aboriginal development of life – and nothing is more likely. Perhaps the prevailing doctrine is in nothing placed in greater difficulties than it is with regard to the entozoa, or creatures which live within the bodies of others. These creatures do, and apparently can, live nowhere else than in the interior of other living bodies, where they generally take up their abode in the viscera, but also sometimes in the chambers of the eye, the interior of the brain, the serous sacs, and other places having no communication from without. Some are viviparous, others oviparous. Of the latter it cannot reasonably be supposed that the ova ever pass through the medium of the air, or through the blood-vessels, for they are too heavy for the one transit, and too large for the other. Of the former, it cannot be conceived how they pass into young animals – certainly not by communication from the parent, for it has often been found that entozoa do not appear in certain generations, and some of peculiar and noted character have only appeared at rare intervals, and in very extraordinary circumstances. A candid view of the less popular doctrine, as to the origin of this humble form of life, is taken by a distinguished living naturalist. "To explain the beginning of these worms within the human body, on the common doctrine that all created beings proceed from their likes, or a primordial egg, is so difficult, that the moderns have been driven to speculate, as our fathers did, on their spontaneous birth; but they have received the hypothesis with some modification. Thus it is not from putrefaction or fermentation that the entozoa are born, for both of these processes are rather fatal to their existence, but from the aggregation and fit apposition of matter which is already organized, or has been thrown from organized surfaces.... Their origin in this manner is not more wonderful or more inexplicable than that of many of the inferior animals from sections of themselves.... Particles of matter fitted by digestion, and their transmission through a living body, for immediate assimilation with it, or flakes of lymph detached from surfaces already organized, seem neither to exceed nor fall below that simplicity of structure which favours this wonderful development; and the supposition that, like morsels of a planaria, they may also, when retained in contact with living parts, and in other favourable circumstances, continue to live and be gradually changed into creatures of analogous conformation, is surely not so absurd as to be brought into comparison with the Metamorphoses of Ovid.... We think the hypothesis is also supported in some degree by the fact, that the origin of the entozoa is favoured by all causes which tend to disturb the equality between the secerning and absorbent systems."* Here particles of organized matter are suggested as the germinal origin of distinct and fully organized animals, many of which have a highly developed reproductive system. How near such particles must be to the inorganic form of matter may be judged from what has been said within the last few pages. If, then, this view of the production of entozoa be received, it must be held as in no small degree favourable to the general doctrine of an organic creation by law. ....

[184-187] The Eternal One has arranged for everything beforehand, and trusted all to the operation of the laws of his appointment, himself being ever present in all things. We can even conceive that man, in his many doings upon the surface of the earth, may occasionally, without his being aware of it, or otherwise, act as an instrument in preparing the association of conditions under which the creative laws work; and perhaps some instances of his having acted as such an instrument have actually occurred in our own time.

I allude, of course, to the experiments conducted a few years ago by Mr. Crosse, which seemed to result in the production of a heretofore unknown species of insect in considerable numbers. Various causes have prevented these experiments and their results from receiving candid treatment, but they may perhaps be yet found to have opened up a new and most interesting chapter of nature's mysteries. Mr. Crosse was pursuing some experiments in crystallization, causing a powerful voltaic battery to operate upon a saturated solution of silicate of potash, when the insects unexpectedly made their appearance. He afterwards tried nitrate of copper, which is a deadly poison, and from that fluid also did live insects emerge. Discouraged by the reception of his experiments, Mr. Crosse soon discontinued them; but they were some years after pursued by Mr. Weekes, of Sandwich, with precisely the same results. This gentleman, besides trying the first of the above substances, employed ferro-cyanet of potash, on account of its containing a larger proportion of carbon, the principal element of organic bodies; and from

* Article "Zoophytes," Encyclopaedia Britannica, 7th edition.

this substance the insects were produced *in increased numbers.* A few weeks sufficed for this experiment, with the powerful battery of Mr. Crosse; but the first attempts of Mr. Weekes required about eleven months, a ground of presumption in itself that the electricity was chiefly concerned in the phenomenon. The changes undergone by the fluid operated upon, were in both cases remarkable, and nearly alike. In Mr. Weekes' apparatus, the silicate of potash became first turbid, then of a milky appearance; round the negative wire of the battery, dipped into the fluid, there gathered a quantity of *gelatinous matter,* a part of the process of considerable importance, considering that gelatin is one of the *proximate principles,* or first compounds, of which animal bodies are formed. From this matter Mr. Weekes observed one of the insects in the very act of emerging, immediately after which, it ascended to the surface of the fluid, and sought concealment in an obscure corner of the apparatus. The insects produced by both experimentalists seem to have been the same, a species of acarus, minute and semi-transparent, and furnished with long bristles, which can only be seen by the aid of the microscope. It is worthy of remark, that some of these insects, soon after their existence had commenced, were found to be likely to extend their species. They were sometimes observed to go back to the fluid to feed, and occasionally they devoured each other.* . . . .

[202-205] The tendency of all these illustrations is to make us look to *development* as the principle which has been immediately concerned in the peopling of this globe, a process extending over a vast space of time, but which is nevertheless connected in character with the briefer process by which an individual being is evoked from a simple germ. What mystery is there here – and how shall I proceed to enunciate the conception which I have ventured to form of what may prove to be its proper solution! It is an idea by no means calculated to impress by its greatness, or to puzzle by its profoundness. It is an idea more marked by simplicity than perhaps any other of those which have explained the great secrets of nature. But in this lies, perhaps, one of its strongest claims to the faith of mankind.

The whole train of animated beings, from the simplest and oldest up to the highest and most recent, are, then, to be regarded as a series of *advances of the principle of development,* which have depended upon external physical circumstances, to which the resulting animals are appropriate. I contemplate the whole phenomena as having been in the

* See a pamphlet circulated by Mr. Weekes, in 1842.

first place arranged in the counsels of Divine Wisdom, to take place, not only upon this sphere, but upon all the others in space, under necessary modifications, and as being carried on, from first to last, here and elsewhere, under immediate favour of the creative will or energy.* The nucleated vesicle, the fundamental form of all organization, we must regard as the meeting-point between the inorganic and the organic – the end of the mineral and beginning of the vegetable and animal kingdoms, which thence start in different directions, but in perfect parallelism and analogy. We have already seen that this nucleated vesicle is itself a type of mature and independent being in the infusory animalcules, as well as the starting point of the foetal progress of every higher individual in creation, both animal and vegetable. We have seen that it is a form of being which electric agency will produce – though not perhaps usher into full life – in albumen, one of those compound elements of animal bodies, of which another (urea) has been made by artificial means. Remembering these things, we are drawn on to the supposition, that the first step in the creation of life upon this planet was *a chemico-electric operation, by which simple germinal vesicles were produced.* This is so much, but what were the next steps? Let a common vegetable infusion help us to an answer. There, as we have seen, simple forms are produced at first, but afterwards they become more complicated, until at length the life-producing powers of the infusion are exhausted. Are we to presume that, in this case, the simple engender the complicated? Undoubtedly, this would not be more wonderful as a natural process than one which we never think of wondering at, because familiar to us – namely, that in the gestation of the mammals, the animalcule-like ovum of a few days is the parent, in a sense, of the chick-like form of a few weeks, and that in all the subsequent stages – fish, reptile, &c. – the one may, with scarcely a metaphor, be said to be the progenitor of the other. I suggest, then, as an hypothesis already countenanced by much that is ascertained, and likely to be further sanctioned by much that remains to be known, that the first step was *an advance under favour of peculiar conditions, from the simplest forms of being, to the next more complicated, and this through the medium of the ordinary process of generation.* ....

* When I formed this idea, I was not aware of one which seems faintly to foreshadow it – namely, Socrates's doctrine, afterwards dilated on by Plato, that "previous to the existence of the world, and beyond its present limits, there existed certain archetypes, the embodiment (if we may use such a word) of general ideas; and that these archetypes were models, in imitation of which all particular beings were created."

[219-223] Perhaps even the transition from species to species does still take place in some of the obscurer fields of creation, or under extraordinary casualties, though science professes to have no such facts on record. It is here to be remarked, that such facts might often happen, and yet no record be taken of them, for so strong is the prepossession for the doctrine of invariable like-production, that such circumstances, on occurring, would be almost sure to be explained away on some other supposition, or, if presented, would be disbelieved and neglected. Science, therefore, has no such facts, for the very same reason that some small sects are said to have no discreditable members – namely, that they do not receive such persons, and extrude all who begin to verge upon the character. There are, nevertheless, some facts which have chanced to be reported without any reference to this hypothesis, and which it seems extremely difficult to explain satisfactorily upon any other. One of these has already been mentioned – a progression in the forms of the animalcules in a vegetable infusion from the simpler to the more complicated, a sort of microcosm, representing the whole history of the progress of animal creation as displayed by geology. Another is given in the history of the Acarus Crossii, which may be only the ultimate stage of a series of similar transformations effected by electric agency in the solution subjected to it. There is, however, one direct case of a translation of species, which has been presented with a respectable amount of authority.* It appears that, whenever oats sown at the usual time are kept cropped down during summer and autumn, and allowed to remain over the winter, a thin crop of rye is the harvest presented at the close of the ensuing summer. This experiment has been tried repeatedly, with but one result; invariably the *secale cereale* is the crop reaped where the *avena sativa,* a recognised different species, was sown. Now it will not satisfy a strict inquirer to be told that the seeds of the rye were latent in the ground and only superseded the dead product of the oats; for if any such fact were in the case, why should the usurping grain be always rye? Perhaps those curious facts which have been stated with regard to forests of one kind of trees, when burnt down, being succeeded (without planting) by other kinds, may yet be found most explicable, as this is, upon the hypothesis of a progression of species which takes place under certain favouring conditions, now apparently of comparatively rare occurrence. The case of the oats is the more valuable, as bearing upon the

* See an article by Dr. Weissenborn, in the New Series of "Magazine of Natural History," vol. i. p. 574.

suggestion as to a protraction of the gestation at a particular part of its course. Here, the generative process is, by the simple mode of cropping down, kept up for a whole year beyond its usual term. The type is thus allowed to advance, and what was oats becomes rye.

The idea, then, which I form of the progress of organic life upon the globe – and the hypothesis is applicable to all similar theaters of vital being – is, *that the simplest and most primitive type, under a law to which that of like-production is subordinate, gave birth to the type next above it, that this again produced the next higher, and so on to the very highest,* the stages of advance being in all cases very small – namely, from one species only to another; so that the phenomenon has always been of a simple and modest character. Whether the whole of any species was at once translated forward, or only a few parents were employed to give birth to the new type, must remain undetermined; but, supposing that the former was the case, we must presume that the moves along the line or lines were simultaneous, so that the place vacated by one species was immediately taken by the next in succession, and so on back to the first, for the supply of which the formation of a new germinal vesicle out of inorganic matter was alone necessary. Thus, the production of new forms, as shewn in the pages of the geological record, has never been anything more than a new stage of progress in gestation, an event as simply natural, and attended as little by any circumstances of a wonderful or startling kind, as the silent advance of an ordinary mother from one week to another of her pregnancy. Yet, be it remembered, the whole phenomena are, in another point of view, wonders of the highest kind, for in each of them we have to trace the effect of an Almighty Will which had arranged the whole in such harmony with external physical circumstances, that both were developed in parallel steps – and probably this development upon our planet is but a sample of what has taken place, through the same cause, in all the other countless theatres of being which are suspended in space. ....

[230-231] Early in this century, M. Lamarck, a naturalist of the highest character, suggested an hypothesis of organic progress which deservedly incurred much ridicule, although it contained a glimmer of the truth. He surmised, and endeavoured, with a great deal of ingenuity, to prove, that one being advanced in the course of generations to another, in consequence merely of its experience of wants calling for the exercise of its faculties in a particular direction, by which exercise

new developments of organs took place, ending in variations sufficient to constitute a new species. Thus he thought that a bird would be driven by necessity to seek its food in the water, and that, in its efforts to swim, the outstretching of its claws would lead to the expansion of the intermediate membranes, and it would thus become web-footed. Now it is possible that wants and the exercise of faculties have entered in some manner into the production of the phenomena which we have been considering; but certainly not in the way suggested by Lamarck, whose whole notion is obviously so inadequate to account for the rise of the organic kingdoms, that we only can place it with pity among the follies of the wise. Had the laws of organic development been known in his time, his theory might have been of a more imposing kind. It is upon these that the present hypothesis is mainly founded. I take existing natural means, and shew them to have been capable of producing all the existing organisms, with the simple and easily conceivable aid of a higher generative law, which we perhaps still see operating upon a limited scale. I also go beyond the French philosopher to a very important point, the original Divine conception of all the forms of being which these natural laws were only instruments in working out and realizing.

•

1855

Alfred Russel Wallace
The Introduction of New Species

[Alfred Russel Wallace, "On the Law Which Has Regulated the Introduction of New Species," in *The Annals and Magazine of Natural History, Including Zoology, Botany, and Geology,* 16, 2 (1855), 184-196]

EVERY NATURALIST who has directed his attention to the subject of the geographical distribution of animals and plants, must have been interested in the singular facts which it presents. Many of these facts are quite different from what would have been anticipated, and have hitherto been considered as highly curious, but quite inexplicable. None of the explanations attempted from the time of Linnaeus are now considered at all satisfactory; none of them have given a cause sufficient to account for the facts known at the time, or comprehensive enough to include all the new facts which have since been, and are daily being added. Of late years, however, a great light has been thrown upon the subject by geological investigations, which have shown that the present state of the earth, and the organisms now inhabiting it, are but the last stage of a long and uninterrupted series of changes which it has undergone, and consequently, that to endeavour to explain and account for its present condition without any reference to those changes (as has frequently been done) must lead to very imperfect and erroneous conclusions.

The facts proved by geology are briefly these:—That during an immense, but unknown period, the surface of the earth has undergone successive changes; land has sunk beneath the ocean, while fresh land has risen up from it; mountain chains have been elevated; islands have been formed into continents, and continents submerged till they have become islands; and these changes have taken place, not once merely, but perhaps hundreds, perhaps thousands of times: – That all these operations have been more or less continuous, but unequal in their progress, and during the whole series the organic life of the earth has undergone a corresponding alteration. This alteration also has been gradual, but complete; after a certain interval not a single species existing which had lived at the commencement of the period. This complete renewal of the forms of life also appears to have occurred several times: – That from the last of the Geological epochs to the present or Historical epoch, the change of organic life has been gradual: the first appearance of animals now existing can in many cases be traced, their numbers gradually increasing in the more recent formations, while other species continually die out and disappear, so that the present condition of the organic world is clearly derived by a natural process of gradual extinction and creation of species from that of the latest geological periods. We may therefore safely infer a like gradation and natural sequence from one geological epoch to another.

Now, taking this as a fair statement of the results of geological inquiry, we see that the present geographical distribution of life upon the earth must be the result of all the previous changes, both of the surface of the earth itself and of its inhabitants. Many causes no doubt have operated of which we must ever remain in ignorance, and we may therefore expect to find many details very difficult of explanation, and in attempting to give one, must allow ourselves to call into our service geological changes which it is highly probable may have occurred, though we have no direct evidence of their individual operation.

The great increase of our knowledge within the last twenty years, both of the present and past history of the organic world, has accumulated a body of facts which should afford a sufficient foundation for a comprehensive law embracing and explaining them all, and giving a direction to new researches. It is about ten years since the idea of such a law suggested itself to the writer of this paper, and he has since taken every opportunity of testing it by all the newly ascertained facts with which he has become acquainted, or has been able to observe himself. These have all served to convince him of the correctness of his hypothesis. Fully to enter into such a subject would occupy much space, and it is only in consequence of some views having been lately promulgated, he believes in a wrong direction, that he now ventures to present his ideas to the public, with only such obvious illustrations of the arguments and results as occur to him in a place far removed from all means of reference and exact information.

The following propositions in Organic Geography and Geology give the main facts on which the hypothesis is founded.

Geography.

1. Large groups, such as classes and orders, are generally spread over the whole earth, while smaller ones, such as families and genera, are frequently confined to one portion, often to a very limited district.

2. In widely distributed families the genera are often limited in range; in widely distributed genera, well-marked groups of species are peculiar to each geographical district.

3. When a group is confined to one district, and is rich in species, it is almost invariably the case that the most closely allied species are found in the same locality or in closely adjoining localities, and that therefore the natural sequence of the species by affinity is also geographical.

4. In countries of a similar climate, but separated by a wide sea or lofty mountains, the families, genera and species of the one are often represented by closely allied families, genera and species peculiar to the other.

Geology.

5. The distribution of the organic world in time is very similar to its present distribution in space.

6. Most of the larger and some small groups extend through several geological periods.

7. In each period, however, there are peculiar groups, found nowhere else, and extending through one or several formations.

8. Species of one genus, or genera of one family occurring in the same geological time are more closely allied than those separated in time.

9. As generally in geography no species or genus occurs in two very distant localities without being also found in intermediate places, so in geology the life of a species or genus has not been interrupted. In other words, no group or species has come into existence twice.

10. The following law may be deduced from these facts:— *Every*

species has come into existence coincident both in space and time with a pre-existing closely allied species.

This law agrees with, explains and illustrates all the facts connected with the following branches of the subject:– 1st. The system of natural affinities. 2nd. The distribution of animals and plants in space. 3rd. The same in time, including all the phaenomena of representative groups, and those which Professor Forbes supposed to manifest polarity. 4th. The phaenomena of rudimentary organs. We will briefly endeavour to show its bearing upon each of these.

If the law above enunciated be true, it follows that the natural series of affinities will also represent the order in which the several species came into existence, each one having had for its immediate antitype a closely allied species existing at the time of its origin. It is evidently possible that two or three distinct species may have had a common antitype, and that each of these may again have become the antitypes from which other closely allied species were created. The effect of this would be, that so long as each species has had but one new species formed on its model, the line of affinities will be simple, and may be represented by placing the several species in direct succession in a straight line. But if two or more species have been independently formed on the plan of a common antitype, then the series of affinities will be compound, and can only be represented by a forked or many-branched line. Now, all attempts at a Natural classification and arrangement of organic beings show, that both these plans have obtained in creation. Sometimes the series of affinities can be well represented for a space by a direct progression from species to species or from group to group, but it is generally found impossible so to continue. There constantly occur two or more modifications of an organ or modifications of two distinct organs, leading us on to two distinct series of species, which at length differ so much from each other as to form distinct genera or families. These are the parallel series or representative groups of naturalists, and they often occur in different countries, or are found fossil in different formations. They are said to have an analogy to each other when they are so far removed from their common antitype as to differ in many important points of structure, while they still preserve a family resemblance. We thus see how difficult it is to determine in every case whether a given relation is an analogy or an affinity, for it is evident that as we go back along the parallel or divergent series, towards the common antitype, the analogy which existed between the two groups becomes an affinity. We are also made aware of the difficulty of arriving at a true classification, even in a small and perfect group; – in the actual state of nature it is almost impossible, the species being so numerous and the modifications of form and structure so varied, arising probably from the immense number of species which have served as antitypes for the existing species, and thus produced a complicated branching of the lines of affinity, as intricate as the twigs of a gnarled oak or the vascular system of the human body. Again, if we consider that we have only fragments of this vast system, the stem and main branches being represented by extinct species of which we have no knowledge, while a vast mass of limbs and boughs and minute twigs and scattered leaves is what we have to place in order, and determine the true position each originally occupied with regard to the others, the whole difficulty of the true Natural System of classification becomes apparent to us.

We shall thus find ourselves obliged to reject all those systems of classification which arrange species or groups in circles, as well as those which fix a definite number for the divisions of each group. The latter class have been very generally rejected by naturalists, as contrary to nature, notwithstanding the ability with which they have been advocated; but the circular system of affinities seems to have obtained a deeper hold, many eminent naturalists having to some extent adopted it. We have, however, never been able to find a case in which the circle has been closed by a direct and close affinity. In most cases a palpable analogy has been substituted, in others the affinity is very obscure or altogether doubtful. The complicated branching of the lines of affinities in extensive groups must also afford great facilities for giving a show of probability to any such purely artificial arrangements. Their death-blow was given by the admirable paper of the lamented Mr. Strickland, published in the 'Annals of Natural History,' in which he so clearly showed the true synthetical method of discovering the Natural System.

If we now consider the geographical distribution of animals and plants upon the earth, we shall find all the facts beautifully in accordance with, and readily explained by, the present hypothesis. A country having species, genera, and whole families peculiar to it, will be the necessary result of its having been isolated for a long period, sufficient for many series of species to have been created on the type of pre-existing ones, which, as well as many of the earlier-formed species, have become extinct, and thus made the groups appear isolated. If in any case the antitype had an extensive range, two or more groups of species might have been formed, each varying from it in a different manner, and thus producing several representative or analogous groups. The *Sylviadae* of Europe and the *Sylvicolidae* of North America, the *Heliconidae* of South America and the *Euploeas* of the East, the group of *Trogons* inhabiting Asia, and that peculiar to South America, are examples that may be accounted for in this manner.

Such phaenomena as are exhibited by the Galapagos Islands, which contain little groups of plants and animals peculiar to themselves, but most nearly allied to those of South America, have not hitherto received any, even a conjectural explanation. The Galapagos are a volcanic group of high antiquity, and have probably never been more closely connected with the continent than they are at present. They must have been first peopled, like other newly-formed islands, by the action of winds and currents, and at a period sufficiently remote to have had the original species die out, and the modified prototypes only remain. In the same way we can account for the separate islands having each their peculiar species, either on the supposition that the same original emigration peopled the whole of the islands with the same species from which differently modified prototypes were created, or that the islands were successively peopled from each other, but that new species have been created in each on the plan of the pre-existing ones. St. Helena is a similar case of a very ancient island having obtained an entirely peculiar, though limited, flora. On the other hand, no example is known of an island which can be proved geologically to be of very recent origin (late in the Tertiary, for instance), and yet possesses generic or family groups, or even many species peculiar to itself.

When a range of mountains has attained a great elevation, and has so remained during a long geological period, the species of the two sides at and near their bases will be often very different, representative species of some genera occurring, and even whole genera being peculiar to one side only, as is remarkably seen in the case of the Andes and Rocky Mountains. A similar phaenomenon occurs when an island has been separated from a continent at a very early period. The shallow sea between the Peninsula of Malacca, Java, Sumatra and Borneo was probably a continent or large island at an early epoch, and may have become submerged as the volcanic ranges of Java and Sumatra were elevated. The organic results we see in the very considerable number of species of animals common to some or all of these countries, while at the same time a number of closely allied representative species exist peculiar to each, showing that a considerable period has elapsed since their separation. The facts of geographical distribution and of geology may thus mutually explain each other in doubtful cases, should the principles here advocated be clearly established.

In all those cases in which an island has been separated from a continent, or raised by volcanic or coralline action from the sea, or in which a mountain-chain has been elevated, in a recent geological epoch, the phaenomena of peculiar groups or even of single representative species will not exist. Our own island is an example of this, its separation from the continent being geologically very recent, and we have consequently scarcely a species which is peculiar to it; while the Alpine range, one of the most recent mountain elevations, separates faunas and floras which scarcely differ more than may be due to climate and latitude alone.

The series of facts alluded to in Proposition 3, of closely allied species in rich groups being found geographically near each other, is most striking and important. Mr. Lovell Reeve has well exemplified it in his able and interesting paper on the Distribution of the *Bulimi.* It is also seen in the Humming-birds and Toucans, little groups of two or three closely allied species being often found in the same or closely adjoining districts, as we have had the good fortune of personally verifying. Fishes give evidence of a similar kind: each great river has its peculiar genera, and in more extensive genera its groups of closely allied species. But it is the same throughout Nature; every class and order of animals will contribute similar facts. Hitherto no attempt has been made to explain these singular phaenomena, or to show how they have arisen. Why are the genera of Palms and of Orchids in almost every case confined to one hemisphere? Why are the closely allied species of brown-backed Trogons all found in the East, and the green-backed in the West? Why are the Macaws and the Cockatoos similarly restricted? Insects furnish a countless number of analogous examples; – the *Goliathi* of Africa, the *Ornithopterae* of the Indian islands, the *Heloconidae* of South America, the *Danaidae* of the East, and in all, the most closely allied species found in geographical proximity. The question forces itself upon every thinking mind, – why are these things so? They could not be as they are, had no law regulated their creation and dispersion. The law here enunciated not merely explains, but necessitates the facts we see to exist, while the vast and long-continued geological changes of the earth readily account for the exceptions and apparent discrepancies that here and there occur. The writer's object in putting forward his views in the present imperfect manner is to submit them to the test of other minds, and to be made aware of all the facts supposed to be inconsistent with them. As his hypothesis is one which claims acceptance solely as explaining and connecting facts which exist in nature, he expects facts alone to be brought to disprove it; not *à-priori* arguments against its probability.

The phaenomena of geological distribution are exactly analogous to those of geography. Closely allied species are found associated in the same beds, and the change from species to species appears to have been as gradual in time as in space. Geology, however, furnishes us with positive proof of the extinction and production of species, though it does not inform us how either has taken place. The extinction of species, however, offers but little difficulty, and the *modus operandi* has been well illustrated by Sir C. Lyell in his admirable 'Principles.' Geological changes, however gradual, must occasionally have modified external conditions to such an extent as to have rendered the existence of certain species impossible. The extinction would in most cases be effected by a gradual dying-out, but in some instances there might have been a sudden destruction of a species of limited range. To discover how the extinct species have from time to time been replaced by new ones down to the very latest geological period, is the most difficult, and at the same time the most interesting problem in the natural history of the earth. The present inquiry, which seeks to eliminate from known facts a law which has determined, to a certain degree, what species could and did appear at a given epoch, may, it is hoped, be considered as one step in the right direction towards a complete solution of it.

Much discussion has of late years taken place on the question, whether the succession of life upon the globe has been from a lower to a higher degree of organization? The admitted facts seem to show that there has been a general, but not a detailed progression. Mollusca and Radiata existed before Vertebrata, and the progression from Fishes to Reptiles and Mammalia, and also from the lower mammals to the higher, is indisputable. On the other hand, it is said that the Mollusca and Radiata of the very earliest periods were more highly organized than the great mass of those now existing, and that the very first fishes that have been discovered are by no means the lowest organized of the class. Now it is believed the present hypothesis will harmonize with all these facts, and in a great measure serve to explain them; for though it may appear to some readers essentially a theory of progression, it is in reality only one of gradual change. It is, however, by no means difficult to show that a real progression in the scale of organization is perfectly consistent with all the appearances, and even with apparent retrogression, should such occur.

Returning to the analogy of a branching tree, as the best mode of representing the natural arrangement of species and their successive creation, let us suppose that at an early geological epoch any group (say a class of the Mollusca) has attained to a great richness of species and a high organization. Now let this great branch of allied species, by geological mutations, be completely or partially destroyed. Subsequently a new branch springs from the same trunk, that is to say, new species are successively created, having for their antitypes the same lower organized species which had served as the antitypes for the former group, but which have survived the modified conditions which destroyed it. This new group being subject to these altered conditions, has modifications of structure and organization given to it, and becomes the representative group of the former one in another geological formation. It may, however, happen, that though later in time, the new series of species may never attain to so high a degree of organization as those preceding it, but in its turn become extinct, and give place to yet another modification from the same root, which may be of higher or lower organization, more or less numerous in species, and more or less varied in form and structure than either of those which preceded it. Again, each of these groups may not have become totally extinct, but may have left a few species, the modified prototypes of which have existed in each succeeding period, a faint memorial of their former grandeur and luxuriance. Thus every case of apparent retrogression may be in reality a progress, though an interrupted one: when some monarch of the forest loses a limb, it may be replaced by a feeble and sickly substitute. The foregoing remarks appear to apply to the case of the Mollusca, which, at a very early period, had reached a high organization and a great development of forms and species in the Testaceous Cephalopoda. In each succeeding age modified species and genera replaced the former ones which had become extinct, and as we approach the present aera but few and small representatives of the group remain, while the Gasteropods and Bivalves have acquired an immense preponderance. In the long series of changes the earth has undergone, the process of peopling it with organic beings has been continually going on, and whenever any of the higher groups have become nearly or quite extinct, the lower forms which have better resisted the modified physical conditions have served as the antitypes on which to found the new races. In this manner alone, it is believed, can the representative groups at successive periods, and the risings and fallings in the scale of organization, be in every case explained.

The hypothesis of polarity, recently put forward by Professor Edward Forbes* to account for the abundance of generic forms at a very early period and at present, while in the intermediate epochs there is a gradual diminution and impoverishment, till the minimum occurred at the confines of the Palaeozoic and Secondary epochs, appears to us quite unnecessary, as the facts may be readily accounted for on the principles already laid down. Between the Palaeozoic and Neozoic periods of Professor Forbes, there is scarcely a species in common, and the greater part of the genera and families also disappear to be replaced by new ones. It is almost universally admitted that such a change in the organic world must have occupied a vast period of time. Of this interval we have no record; probably because the whole area of the early formations now exposed to our researches was elevated at the end of the Palaeozoic period, and remained so through the interval required for the organic changes which resulted in the fauna and flora of the Secondary period. The records of this interval are buried beneath [the] ocean which covers three-fourths of the globe. Now it appears highly probable that a long period of quiescence or stability in the physical conditions of a district would be most favourable to the existence of organic life in the greatest abundance, both as regards individuals and also as to variety of species and generic groups, just as we now find that the places best adapted to the rapid growth and increase of individuals also contain the greatest profusion of species and the greatest variety of forms,– the tropics in comparison with the temperate and arctic regions. On the other hand, it seems no less probable that a change in the physical conditions of a district, even small in amount if rapid, or even gradual if to a great amount, would be highly unfavourable to the existence of individuals, might cause the extinction of many species, and would probably be equally unfavourable to the creation of new ones. In this too we may find an analogy with the present state of our earth, for it has been shown to be the violent extremes and rapid changes of physical conditions, rather than the actual mean state in the

---

*Since the above was written, the author has heard with sincere regret of the death of this eminent naturalist, from whom so much important work was expected. His remarks on the present paper,– a subject on which no man was more competent to decide, – were looked for with the greatest interest. Who shall supply his place?

temperate and frigid zones, which renders them less prolific than the tropical regions, as exemplified by the great distance beyond the tropics to which tropical forms penetrate when the climate is equable, and also by the richness in species and forms of tropical mountain regions which principally differ from the temperate zone in the uniformity of their climate. However this may be, it seems a fair assumption that during a period of geological repose the new species which we know to have been created would have appeared, that the creations would then exceed in number the extinctions, and therefore the number of species would increase. In a period of geological activity, on the other hand, it seems probable that the extinctions might exceed the creations, and the number of species consequently diminish. That such effects did take place in connexion with the causes to which we have imputed them, is shown in the case of the Coal formation, the faults and contortions of which show a period of great activity and violent convulsions, and it is in the formation immediately succeeding this that the poverty of forms of life is most apparent. We have then only to suppose a long period of somewhat similar action during the vast unknown interval at the termination of the Palaeozoic period, and then a decreasing violence or rapidity through the Secondary period, to allow for the gradual repopulation of the earth with varied forms, and the whole of the facts are explained. We thus have a clue to the increase of the forms of life during certain periods, and their decrease during others, without recourse to any causes but those we know to have existed, and to effects fairly deducible from them. The precise manner in which the geological changes of the early formations were effected is so extremely obscure, that when we can explain important facts by a retardation at one time and an acceleration at another of a process which we know from its nature and from observation to have been unequal, – a cause so simple may surely be preferred to one so obscure and hypothetical as polarity.

I would also venture to suggest some reasons against the very nature of the theory of Professor Forbes. Our knowledge of the organic world during any geological epoch is necessarily very imperfect. Looking at the vast numbers of species and groups that have been discovered by geologists, this may be doubted; but we should compare their numbers not merely with those that now exist upon the earth, but with a far larger amount*. We have no reason for believing that the number of

(*See on this subject a paper by Professor Agassiz in the 'Annals' for November 1854.—Ed.)

species on the earth at any former period was much less than at present; at all events the aquatic portion, with which geologists have most acquaintance, was probably often as great or greater. Now we know that there have been many complete changes of species; new sets of organisms have many times been introduced in place of old ones which have become extinct, so that the total amount which have existed on the earth from the earliest geological period must have borne about the same proportion to those now living, as the whole human race who have lived and died upon the earth, to the population at the present time. Again, at each epoch, the whole earth was no doubt, as now, more or less the theatre of life, and as the successive generations of each species died, their exuviae and preservable parts would be deposited over every portion of the then existing seas and oceans, which we have reason for supposing to have been more, rather than less, extensive than at present. In order then to understand our possible knowledge of the early world and its inhabitants, we must compare, not the area of the whole field of our geological researches with the earth's surface, but the area of the examined portion of each formation separately with the whole earth. For example, during the Silurian period all the earth was Silurian, and animals were living and dying, and depositing their remains more or less over the whole area of the globe, and they were probably (the species at least) nearly as varied in different latitudes and longitudes as at present. What proportion do the Silurian districts bear to the whole surface of the globe, land and sea (for far more extensive Silurian districts probably exist beneath the ocean than above it), and what portion of the known Silurian districts has been actually examined for fossils? Would the area of rock actually laid open to the eye be the thousandth or the ten-thousandth part of the earth's surface? Ask the same question with regard to the Oolite or the Chalk, or even to particular beds of these when they differ considerably in their fossils, and you may then get some notion of how small a portion of the whole we know.

But yet more important is the probability, nay, almost the certainty, that whole formations containing the records of vast geological periods are entirely buried beneath the ocean, and for ever beyond our reach. Most of the gaps in the geological series may thus be filled up, and vast numbers of unknown and unimaginable animals, which might help to elucidate the affinities of the numerous isolated groups which are a perpetual puzzle to the zoologist, may there be buried, till future revolutions may raise them in their turn above the waters, to afford materials for the study of whatever race of intelligent beings may then have succeeded us. These considerations must lead us to the conclusion, that our knowledge of the whole series of the former inhabitants of the earth is necessarily most imperfect and fragmentary, – as much so as our knowledge of the present organic world would be, were we forced to make our collections and observations only in spots equally limited in area and in number with those actually laid open for the collection of fossils. Now, the hypothesis of Professor Forbes is essentially one that assumes to a great extent the *completeness* of our knowledge of the *whole series* of organic beings which have existed on the earth. This appears to be a fatal objection to it, independently of all other considerations. It may be said that the same objections exist against every theory on such a subject, but this is not necessarily the case. The hypothesis put forward in this paper depends in no degree upon the completeness of our knowledge of the former condition of the organic world, but takes what facts we have as fragments of a vast whole, and deduces from them something of the nature and proportions of that whole which we can never know in detail. It is founded upon isolated groups of facts, recognizes their isolation, and endeavours to deduce from them the nature of the intervening portions.

Another important series of facts, quite in accordance with, and even necessary deductions from, the law now developed, are those of *rudimentary organs.* That these really do exist, and in most cases have no special function in the animal oeconomy, is admitted by the first authorities in comparative anatomy. The minute limbs hidden beneath the skin in many of the snake-like lizards, the anal hooks of the boa constrictor, the complete series of jointed finger-bones in the paddle of the Manatus and whale, are a few of the most familiar instances. In botany a similar class of facts has been long recognized. Abortive stamens, rudimentary floral envelopes and undeveloped carpels, are of the most frequent occurrence. To every thoughtful naturalist the question must arise, What are these for? What have they to do with the great laws of creation? Do they not teach us something of the system of Nature? If each species has been created independently, and without any necessary relations with pre-existing species, what do these rudiments, these apparent imperfections mean? There must be a cause for them; they must be the necessary results of some great natural law. Now, if, as it has been endeavoured to be shown, the great law which has regulated the peopling of the earth with animal and vegetable life is, that every change shall be gradual; that no new creature shall be formed widely differing from anything before existing; that in this, as in everything else in Nature, there shall be gradation and harmony, – then these rudimentary organs are necessary, and are an essential part of the system of Nature. Ere the higher Vertebrata were formed, for instance, many steps were required, and many organs had to undergo modifications from the rudimental condition in which only they had as yet existed. We still see remaining an antitypal sketch of a wing adapted for flight in the scaly flapper of the penguin, and limbs first concealed beneath the skin, and then weakly protruding from it, were the necessary gradations before others should be formed fully adapted for locomotion. Many more of these modifications should we behold, and more complete series of them, had we a view of all the forms which have ceased to live. The great gaps that exist between fishes, reptiles, birds and mammals would then, no doubt, be softened down by intermediate groups, and the whole organic world would be seen to be an unbroken and harmonious system.

It has now been shown, though most briefly and imperfectly, how the law that *"Every species has come into existence coincident both in time and space with a pre-existing closely allied species,"* connects together and renders intelligible a vast number of independent and hitherto unexplained facts. The natural system of arrangement of organic beings, their geographical distribution, their geological sequence, the phaenomena of representative and substituted groups in all their modifications, and the most singular peculiarities of anatomical structure, are all explained and illustrated by it, in perfect accordance with the vast mass of facts which the researches of modern naturalists have brought together, and, it is believed, not materially opposed to any of them. It also claims a superiority over previous hypotheses, on the ground that it not merely explains, but necessitates what exists. Granted the law, and many of the most important facts in Nature could not have been otherwise, but are almost as necessary deductions from it, as are the elliptic orbits of the planets from the law of gravitation.

•

1858

Alfred Russel Wallace and Charles Darwin
On Species and Varieties

[Charles Darwin and Alfred Russel Wallace, "On the Tendency of Species to Form Varieties; and on the Perpetuation of Varieties and Species by Natural Means of Selection," in *Journal of the Proceedings of the Linnean Society, Zoology,* 3 (1859), 46-62. Read July 1, 1858. Published on Aug. 20, 1858, although the title page of the volume bears the date 1859.]

Charles Darwin

[Extracts from a MS. work on Species]

DE CANDOLLE, in an eloquent passage, has declared that all nature is at war, one organism with another, or with external nature. Seeing the contented face of nature, this may at first well be doubted; but reflection will inevitably prove it to be true. The war, however, is not constant, but recurrent in a slight degree at short periods, and more severely at occasional more distant periods; and hence its effects are easily overlooked. It is the doctrine of Malthus applied in most cases with tenfold force. As in every climate there are seasons, for each of its inhabitants, of greater and less abundance, so all annually breed; and the moral restraint which in some small degree checks the increase of mankind is entirely lost. Even slow-breeding mankind has doubled in twenty-five years; and if he could increase his food with greater ease, he would double in less time. But for animals without artificial means, the amount of food for each species must, *on an average,* be constant, whereas the increase of all organisms tends to be geometrical, and in a vast majority of cases at an enormous ratio. Suppose in a certain spot there are eight pairs of birds, and that *only* four pairs of them annually (including double hatches) rear only four young, and that these go on rearing their young at the same rate, then at the end of seven years (a short life, excluding violent deaths, for any bird) there will be 2048 birds, instead of the original sixteen. As this increase is quite impossible, we must conclude either that birds do not rear nearly half their young, or that the average life of a bird is, from accident, not nearly seven years. Both checks probably concur. The same kind of calculation applied to all plants and animals affords results more or less striking, but in very few instances more striking than in man.

Many practical illustrations of this rapid tendency to increase are on record, among which, during peculiar seasons, are the extraordinary numbers of certain animals; for instance, during the years 1826 to 1828, in La Plata, when from drought some millions of cattle perished, the whole country actually *swarmed* with mice. Now I think it cannot be doubted that during the breeding-season all the mice (with the exception of a few males or females in excess) ordinarily pair, and therefore that this astounding increase during three years must be attributed to a greater number than usual surviving the first year, and then breeding, and so on till the third year, when their numbers were brought down to their usual limits on the return of wet weather. Where man has introduced plants and animals into a new and favourable country, there are many accounts in how surprisingly few years the whole country has become stocked with them. This increase would necessarily stop as soon as the country was fully stocked; and yet we have every reason to believe, from what is known of wild animals, that *all* would pair in the spring. In the majority of cases it is most difficult to imagine where the checks fall – though generally, no doubt, on the seeds, eggs, and young; but when we remember how impossible, even in mankind (so much better known than any other animal), it is to infer from repeated casual observations what the average duration of life is, or to discover the different percentage of deaths to births in different countries, we ought to feel no surprise at our being unable to discover where the check falls in any animal or plant. It should always be remembered, that in most cases the checks are recurrent yearly in a small, regular degree, and in an extreme degree during unusually cold, hot, dry, or wet years, according to the constitution of the being in question. Lighten any check in the least degree, and the geometrical powers of increase in every organism will almost instantly increase the average number of the favoured species. Nature may be compared to a surface on which rest ten thousand sharp wedges touching each other and driven inwards by incessant blows. Fully to realize these views much reflection is requisite. Malthus on man should be studied; and all such cases as those of the mice in La Plata, of the cattle and horses when first turned out in South America, of the birds by our calculation, &c., should be well considered. Reflect on the enormous multiplying power *inherent and annually in action* in all animals; reflect on the countless seeds scattered by a hundred ingenious contrivances, year after year, over the whole face of the land; and yet we have every reason to suppose that the average percentage of each of the inhabitants of a country usually remains constant. Finally, let it be borne in mind that this average number of individuals (the external conditions remaining the same) in each country is kept up by recurrent struggles against other species or against external nature (as on the borders of the Arctic regions, where the cold checks life), and that ordinarily each individual of every species holds its place, either by its own struggle and capacity of acquiring nourishment in some period of its life, from the egg upwards; or by the struggle of its parents (in short-lived organisms, when the main check occurs at longer intervals) with other individuals of the *same* or *different* species.

But let the external conditions of a country alter. If in a small degree, the relative proportions of the inhabitants will in most cases simply be slightly changed; but let the number of inhabitants be small, as on an island, and free access to it from other countries be circumscribed, and let the change of conditions continue progressing (forming new stations), in such a case the original inhabitants must cease to be as perfectly adapted to the changed conditions as they were originally. It has been shown in a former part of this work, that such changes of external conditions would, from their acting on the reproductive system, probably cause the organization of those beings which were most affected to become, as under domestication, plastic. Now, can it be doubted, from the struggle each individual has to obtain subsistence, that any minute variation in structure, habits, or instincts, adapting that individual better to the new conditions, would tell upon its vigour and health? In the struggle it would have a better *chance* of surviving; and those of its offspring which inherited the variation, be it ever so slight, would also have a better *chance*. Yearly more are bred than can survive; the smallest grain in the balance, in the long run, must tell on which death shall fall, and which shall survive. Let this work of selection on the one hand, and death on the other, go on for a thousand generations, who will pretend to affirm that it would produce no effect, when we remember what, in a few years, Bakewell effected in cattle, and Western in sheep, by this identical principle of selection?

To give an imaginary example from changes in progress on an island: – let the organization of a canine animal which preyed chiefly on rabbits, but sometimes on hares, become slightly plastic; let these same changes cause the number of rabbits very slowly to decrease, and the number of hares to increase; the effect of this would be that the fox or dog would be driven to try to catch more hares: his organization, however, being slightly plastic, those individuals with the lightest forms, longest limbs, and best eyesight, let the difference be ever so small, would be slightly favoured, and would tend to live longer, and to survive during that time of the year when food was scarcest; they would also rear more young, which would tend to inherit these slight peculiarities. The less fleet ones would be rigidly destroyed. I can see no more reason to doubt that these causes in a thousand generations would produce a marked effect, and adapt the form of the fox or dog to the catching of hares instead of rabbits, than that greyhounds can be improved by selection and careful breeding. So would it be with plants under similar circumstances. If the number of individuals of a species with plumed seeds could be increased by greater powers of dissemination within its own area (that is, if the check to increase fell chiefly on the seeds), those seeds which were provided with ever so little more down, would in the long run be most disseminated; hence a greater number of seeds thus formed would germinate, and would tend to produce plants inheriting the slightly better-adapted down*.

Besides this natural means of selection, by which those individuals are preserved, whether in their egg, or larval, or mature state, which are best adapted to the place they fill in nature, there is a second agency at work in most unisexual animals, tending to produce the same effect, namely, the struggle of the males for the females. These struggles are generally decided by the law of battle, but in the case of birds, apparently, by the charms of their song, by their beauty or their power of courtship, as in the dancing rock-thrush of Guiana. The most vigorous and healthy males, implying perfect adaptation, must generally gain the victory in their contests. This kind of selection, however, is less rigorous than the other; it does not require the death of the less successful, but gives to them fewer descendants. The struggle falls, moreover, at a time of year when food is generally abundant, and perhaps the effect chiefly produced would be the modification of the

* I can see no more difficulty in this, than in the planter improving his varieties of the cotton plant. — C.D. 1858.

secondary sexual characters, which are not related to the power of obtaining food, or to defence from enemies, but to fighting with or rivalling other males. The result of this struggle amongst the males may be compared in some respects to that produced by those agriculturists who pay less attention to the careful selection of all their young animals, and more to the occasional use of a choice mate.

Charles Darwin
[Letter to Asa Gray]

1. It is wonderful what the principle of selection by man, that is the picking out of individuals with any desired quality, and breeding from them, and again picking out, can do. Even breeders have been astounded at their own results. They can act on differences inappreciable to an uneducated eye. Selection has been *methodically* followed in *Europe* for only the last half century; but it was occasionally, and even in some degree methodically followed in the most ancient times. There must have been also a kind of unconscious selection from a remote period, namely in the preservation of the individual animals (without any thought of their offspring) most useful to each race of man in his particular circumstances. The "roguing," as nurserymen call the destroying of varieties which depart from their type, is a kind of selection. I am convinced that intentional and occasional selection has been the main agent in the production of our domestic races; but however this may be, its great power of modification has been indisputably shown in later times. Selection acts only by the accumulation of slight or greater variations, caused by external conditions, or by the mere fact that in generation the child is not absolutely similar to its parent. Man, by this power of accumulating variations, adapts living beings to his wants – may be said to make the wool of one sheep good for carpets, of another for cloth, &c.

2. Now suppose there were a being who did not judge by mere external appearances, but who could study the whole internal organization, who was never capricious, and should go on selecting for one object during millions of generations; who will say what he might not effect? In nature we have some *slight* variation occasionally in all parts; and I think it can be shown that changed conditions of existence is the main cause of the child not exactly resembling its parents; and in nature geology shows us what changes have taken place, and are taking place. We have almost unlimited time; no one but a practical geologist can

fully appreciate this. Think of the Glacial period, during the whole of which the same species at least of shells have existed; there must have been during this period millions on millions of generations.

3. I think it can be shown that there is such an unerring power at work in *Natural Selection* (the title of my book), which selects exclusively for the good of each organic being. The elder De Candolle, W. Herbert, and Lyell have written excellently on the struggle for life; but even they have not written strongly enough. Reflect that every being (even the elephant) breeds at such a rate, that in a few years, or at most a few centuries, the surface of the earth would not hold the progeny of one pair. I have found it hard constantly to bear in mind that the increase of every single species is checked during some part of its life, or during some shortly recurrent generation. Only a few of those annually born can live to propagate their kind. What a trifling difference must often determine which shall survive, and which perish!

4. Now take the case of a country undergoing some change. This will tend to cause some of its inhabitants to vary slightly – not but that I believe most beings vary at all times enough for selection to act on them. Some of its inhabitants will be exterminated; and the remainder will be exposed to the mutual action of a different set of inhabitants, which I believe to be far more important to the life of each being than mere climate. Considering the infinitely various methods which living beings follow to obtain food by struggling with other organisms, to escape danger at various times of life, to have their eggs or seeds disseminated, &c. &c., I cannot doubt that during millions of generations individuals of a species will be occasionally born with some slight variation, profitable to some part of their economy. Such individuals will have a better chance of surviving, and of propagating their new and slightly different structure; and the modification may be slowly increased by the accumulative action of natural selection to any profitable extent. The variety thus formed will either coexist with, or, more commonly, will exterminate its parent form. An organic being, like the woodpecker or misseltoe, may thus come to be adapted to a score of contingences [contingencies] – natural selection accumulating those slight variations in all parts of its structure, which are in any way useful to it during any part of its life.

5. Multiform difficulties will occur to every one, with respect to this theory. Many can, I think, be satisfactorily answered. *Natura non facit saltum* answers one of the most obvious. The slowness of the change, and only a very few individuals undergoing change at any one time,

answers others. The extreme imperfection of our geological records answers others.

6. Another principle, which may be called the principle of divergence, plays, I believe, an important part in the origin of species. The same spot will support more life if occupied by very diverse forms. We see this in the many generic forms in a square yard of turf, and in the plants or insects on any little uniform islet, belonging almost invariably to as many genera and families as species. We can understand the meaning of this fact amongst the higher animals, whose habits we understand. We know that it has been experimentally shown that a plot of land will yield a greater weight if sown with several species and genera of grasses, than if sown with only two or three species. Now, every organic being, by propagating so rapidly, may be said to be striving its utmost to increase in numbers. So it will be with the offspring of any species after it has become diversified into varieties, or sub-species, or true species. And it follows, I think, from the foregoing facts, that the varying offspring of each species will try (only few will succeed) to seize on as many and as diverse places in the economy of nature as possible. Each new variety or species, when formed, will generally take the place of, and thus exterminate its less well-fitted parent. This I believe to be the origin of the classification and affinities of organic beings at all times; for organic beings always *seem* to branch and sub-branch like the limbs of a tree from a common trunk, the flourishing and diverging twigs destroying the less vigorous – the dead and lost branches rudely representing extinct genera and families.

This sketch is *most* imperfect; but in so short a space I cannot make it better. Your imagination must fill up very wide blanks.

Alfred Russel Wallace
"On the Tendency of Varieties to Depart Indefinitely from the Original Type"

One of the strongest arguments which have been adduced to prove the original and permanent distinctness of species is, that *varieties* produced in a state of domesticity are more or less unstable, and often have a tendency, if left to themselves, to return to the normal form of the parent species; and this instability is considered to be a distinctive peculiarity of all varieties, even of those occurring among wild animals in a state of nature, and to constitute a provision for preserving unchanged the originally created distinct species.

In the absence or scarcity of facts and observations as to *varieties* occurring among wild animals, this argument has had great weight with naturalists, and has led to a very general and somewhat prejudiced belief in the stability of species. Equally general, however, is the belief in what are called "permanent or true varieties," – races of animals which continually propagate their like, but which differ so slightly (although constantly) from some other race, that the one is considered to be a *variety* of the other. Which is the *variety* and which the original *species*, there is generally no means of determining, except in those rare cases in which the one race has been known to produce an offspring unlike itself and resembling the other. This, however, would seem quite incompatible with the "permanent invariability of species," but the difficulty is overcome by assuming that such varieties have strict limits, and can never again vary further from the original type, although they may return to it, which, from the analogy of the domesticated animals, is considered to be highly probable, if not certainly proved.

It will be observed that this argument rests entirely on the assumption, that *varieties* occurring in a state of nature are in all respects analogous to or even identical with those of domestic animals, and are governed by the same laws as regards their permanence or further variation. But it is the object of the present paper to show that this assumption is altogether false, that there is a general principle in nature which will cause many *varieties* to survive the parent species, and to give rise to successive variations departing further and further from the original type, and which also produces, in domesticated animals, the tendency of varieties to return to the parent form.

The life of wild animals is a struggle for existence. The full exertion of all their faculties and all their energies is required to preserve their own existence and provide for that of their infant offspring. The possibility of procuring food during the least favourable seasons, and of escaping the attacks of their most dangerous enemies, are the primary conditions which determine the existence both of individuals and of entire species. These conditions will also determine the population of a species; and by a careful consideration of all the circumstances we may be enabled to comprehend, and in some degree to explain, what at first sight appears so inexplicable – the excessive abundance of some species, while others closely allied to them are very rare.

The general proportion that must obtain between certain groups of animals is readily seen. Large animals cannot be so abundant as small ones; the carnivora must be less numerous than the herbivora; eagles and lions can never be so plentiful as pigeons and antelopes; the wild asses of the Tartarian deserts cannot equal in numbers the horses of the more luxuriant prairies and pampas of America. The greater or less fecundity of an animal is often considered to be one of the chief causes of its abundance or scarcity; but a consideration of the facts will show us that it really has little or nothing to do with the matter. Even the least prolific of animals would increase rapidly if unchecked, whereas it is evident that the animal population of the globe must be stationary, or perhaps, through the influence of man, decreasing. Fluctuations there may be; but permanent increase, except in restricted localities, is almost impossible. For example, our own observation must convince us that birds do not go on increasing every year in a geometrical ratio, as they would do, were there not some powerful check to their natural increase. Very few birds produce less than two young ones each year, while many have six, eight, or ten; four will certainly be below the average; and if we suppose that each pair produce young only four times in their life, that will also be below the average, supposing them not to die either by violence or want of food. Yet at this rate how tremendous would be the increase in a few years from a single pair! A simple calculation will show that in fifteen years each pair of birds would have increased to nearly ten millions! whereas we have no reason to believe that the number of the birds of any country increases at all in fifteen or in one hundred and fifty years. With such powers of increase the population must have reached its limits, and have become stationary, in a very few years after the origin of each species. It is evident, therefore, that each year an immense number of birds must perish – as many in fact as are born; and as on the lowest calculation the progeny are each year twice as numerous as their parents, it follows that, whatever be the average number of individuals existing in any given country, *twice that number must perish annually,* – a striking result, but one which seems at least highly probable, and is perhaps under rather than over the truth. It would therefore appear that, as far as the continuance of the species and the keeping up the average number of individuals are concerned, large broods are superfluous. On the average all above *one* become food for hawks and kites, wild cats and weasels, or perish of cold and hunger as winter comes on. This is strikingly proved by the case of particular species; for we find that their abundance in individuals bears no relation whatever to their fertility in producing offspring. Perhaps the most remarkable instance of an immense bird population is that of the passenger pigeon of the United

States, which lays only one, or at most two eggs, and is said to rear generally but one young one. Why is this bird so extraordinarily abundant, while others producing two or three times as many young are much less plentiful? The explanation is not difficult. The food most congenial to this species, and on which it thrives best, is abundantly distributed over a very extensive region, offering such differences of soil and climate, that in one part or another of the area the supply never fails. The bird is capable of a very rapid and long-continued flight, so that it can pass without fatigue over the whole of the district it inhabits, and as soon as the supply of food begins to fail in one place is able to discover a fresh feeding-ground. This example strikingly shows us that the procuring a constant supply of wholesome food is almost the sole condition requisite for ensuring the rapid increase of a given species, since neither the limited fecundity, nor the unrestrained attacks of birds of prey and of man are here sufficient to check it. In no other birds are these peculiar circumstances so strikingly combined. Either their food is more liable to failure, or they have not sufficient power of wing to search for it over an extensive area, or during some season of the year it becomes very scarce, and less wholesome substitutes have to be found; and thus, though more fertile in offspring, they can never increase beyond the supply of food in the least favourable seasons. Many birds can only exist by migrating, when their food becomes scarce, to regions possessing a milder, or at least a different climate, though, as these migrating birds are seldom excessively abundant, it is evident that the countries they visit are still deficient in a constant and abundant supply of wholesome food. Those whose organization does not permit them to migrate when their food becomes periodically scarce, can never attain a large population. This is probably the reason why woodpeckers are scarce with us, while in the tropics they are among the most abundant of solitary birds. Thus the house sparrow is more abundant than the redbreast, because its food is more constant and plentiful, – seeds of grasses being preserved during the winter, and our farm-yards and stubble-fields furnishing an almost inexhaustible supply. Why, as a general rule, are aquatic, and especially sea birds, very numerous in individuals? Not because they are more prolific than others, generally the contrary; but because their food never fails, the sea-shores and river-banks daily swarming with a fresh supply of small mollusca and crustacea. Exactly the same laws will apply to mammals. Wild cats are prolific and have few enemies; why then are they never as abundant as rabbits? The only intelligible answer is, that their supply of food is more precarious. It appears evident, therefore, that so long as a country remains physically unchanged, the numbers of its animal population cannot materially increase. If one species does so, some others requiring the same kind of food must diminish in proportion. The numbers that die annually must be immense; and as the individual existence of each animal depends upon itself, those that die must be the weakest – the very young, the aged, and the diseased, – while those that prolong their existence can only be the most perfect in health and vigour – those who are best able to obtain food regularly, and avoid their numerous enemies. It is, as we commenced by remarking, "a struggle for existence," in which the weakest and least perfectly organized must always succumb.

Now it is clear that what takes place among the individuals of a species must also occur among the several allied species of a group, – viz. that those which are best adapted to obtain a regular supply of food, and to defend themselves against the attacks of their enemies and the vicissitudes of the seasons, must necessarily obtain and preserve a superiority in population; while those species which from some defect of power or organization are the least capable of counteracting the vicissitudes of food, supply, &c., must diminish in numbers, and, in extreme cases, become altogether extinct. Between these extremes the species will present various degrees of capacity for ensuring the means of preserving life; and it is thus we account for the abundance or rarity of species. Our ignorance will generally prevent us from accurately tracing the effects to their causes; but could we become perfectly acquainted with the organization and habits of the various species of animals, and could we measure the capacity of each for performing the different acts necessary to its safety and existence under all the varying circumstances by which it is surrounded, we might be able even to calculate the proportionate abundance of individuals which is the necessary result.

If now we have succeeded in establishing these two points – 1st, *that the animal population of a country is generally stationary, being kept down by a periodical deficiency of food, and other checks*; and, 2nd, *that the comparative abundance or scarcity of the individuals of the several species is entirely due to their organization and resulting habits, which, rendering it more difficult to procure a regular supply of food and to provide for their personal safety in some cases than in others, can only be balanced by a difference in the population which have to exist in a given area* – we shall be in a condition to proceed to the consideration of *varieties,* to which the preceding remarks have a direct and very important application.

Most or perhaps all the variations from the typical form of a species must have some definite effect, however slight, on the habits or capacities of the individuals. Even a change of colour might, by rendering them more or less distinguishable, affect their safety; a greater or less development of hair might modify their habits. More important changes, such as an increase in the power or dimensions of the limbs or any of the external organs, would more or less affect their mode of procuring food or the range of country which they inhabit. It is also evident that most changes would affect, either favourably or adversely, the powers of prolonging existence. An antelope with shorter or weaker legs must necessarily suffer more from the attacks of the feline carnivora; the passenger pigeon with less powerful wings would sooner or later be affected in its powers of procuring a regular supply of food; and in both cases the result must necessarily be a diminution of the population of the modified species. If, on the other hand, any species should produce a variety having slightly increased powers of preserving existence, that variety must inevitably in time acquire a superiority in numbers. These results must follow as surely as old age, intemperance, or scarcity of food produce an increased mortality. In both cases there may be many individual exceptions; but on the average the rule will invariably be found to hold good. All varieties will therefore fall into two classes – those which under the same conditions would never reach the population of the parent species, and those which would in time obtain and keep a numerical superiority. Now, let some alteration of physical conditions occur in the district – a long period of drought, a destruction of vegetation by locusts, the irruption of some new carnivorous animal seeking "pastures new" – any change in fact tending to render existence more difficult to the species in question, and tasking its utmost powers to avoid complete extermination; it is evident that, of all the individuals composing the species, those forming the least numerous and most feebly organized variety would suffer first, and, were the pressure severe, must soon become extinct. The same causes continuing in action, the parent species would next suffer, would gradually diminish in numbers, and with a recurrence of similar unfavourable conditions might also become extinct. The superior variety would then alone remain, and on a return to favourable circumstances would rapidly increase in numbers and occupy the place of the extinct species and variety.

The *variety* would now have replaced the *species*, of which it would be a more perfectly developed and more highly organized form. It would be in all respects better adapted to secure its safety, and to prolong its individual existence and that of the race. Such a variety *could not* return to the original form; for that form is an inferior one, and could never compete with it for existence. Granted, therefore, a "tendency" to reproduce the original type of the species, still the variety must ever remain preponderant in numbers, and under adverse physical conditions *again alone survive*. But this new, improved, and populous race might itself, in course of time, give rise to new varieties, exhibiting several diverging modifications of form, any of which, tending to increase the facilities for preserving existence, must, by the same general law, in their turn become predominant. Here, then, we have *progression and continued divergence* deduced from the general laws which regulate the existence of animals in a state of nature, and from the undisputed fact that varieties do frequently occur. It is not, however, contended that this result would be invariable; a change of physical conditions in the district might at times materially modify it, rendering the race which had been the most capable of supporting existence under the former conditions now the least so, and even causing the extinction of the newer and, for a time, superior race, while the old or parent species and its first inferior varieties continued to flourish. Variations in unimportant parts might also occur, having no perceptible effect on the life-preserving powers; and the varieties so furnished might run a course parallel with the parent species, either giving rise to further variations or returning to the former type. All we argue for is, that certain varieties have a tendency to maintain their existence longer than the original species, and this tendency must make itself felt; for though the doctrine of chances or averages can never be trusted to on a limited scale, yet, if applied to high numbers, the results come nearer to what theory demands, and, as we approach to an infinity of examples, become strictly accurate. Now the scale on which nature works is so vast – the numbers of individuals and periods of time with which she deals approach so near to infinity, that any cause, however slight, and however liable to be veiled and counteracted by accidental circumstances, must in the end produce its full legitimate results.

Let us now turn to domesticated animals, and inquire how varieties produced among them are affected by the principles here enunciated. The essential difference in the condition of wild and domestic animals is this, – that among the former, their wellbeing and very existence depend upon the full exercise and healthy condition of all their senses and physical powers, whereas, among the latter, these are only partially exercised, and in some cases are absolutely unused. A wild animal has to search, and often to labour, for every mouthful of food – to exercise sight, hearing, and smell in seeking it, and in avoiding dangers, in procuring shelter from the inclemency of the seasons, and in providing for the subsistence and safety of its offspring. There is no muscle of its body that is not called into daily and hourly activity; there is no sense or faculty that is not strengthened by continual exercise. The domestic animal, on the other hand, has food provided for it, is sheltered, and often confined, to guard it against the vicissitudes of the seasons, is carefully secured from the attacks of its natural enemies, and seldom even rears its young without human assistance. Half of its senses and faculties are quite useless; and the other half are but occasionally called into feeble exercise, while even its muscular system is only irregularly called into action.

Now when a variety of such an animal occurs, having increased power or capacity in any organ or sense, such increase is totally useless, is never called into action, and may even exist without the animal ever becoming aware of it. In the wild animal, on the contrary, all its faculties and powers being brought into full action for the necessities of existence, any increase becomes immediately available, is strengthened by exercise, and must even slightly modify the food, the habits, and the whole economy of the race. It creates as it were a new animal, one of superior powers, and which will necessarily increase in numbers and outlive those inferior to it.

Again, in the domesticated animal all variations have an equal chance of continuance; and those which would decidedly render a wild animal unable to compete with its fellows and continue its existence are no disadvantage whatever in a state of domesticity. Our quickly fattening pigs, short-legged sheep, pouter pigeons, and poodle dogs could never have come into existence in a state of nature, because the very first step towards such inferior forms would have led to the rapid extinction of the race; still less could they now exist in competition with their wild allies. The great speed but slight endurance of the race horse, the unwieldy strength of the ploughman's team, would both be useless in a state of nature. If turned wild on the pampas, such animals would probably soon become extinct, or under favourable circumstances might each lose those extreme qualities which would never be called into action, and in a few generations would revert to a common type, which must be that in which the various powers and faculties are so proportioned to each other as to be best adapted to procure food and secure safety, – that in which by the full exercise of every part of his organization the animal can alone continue to live. Domestic varieties, when turned wild, *must* return to something near the type of the original wild stock, *or become altogether extinct.*

We see, then, that no inferences as to varieties in a state of nature can be deduced from the observation of those occurring among domestic animals. The two are so much opposed to each other in every circumstance of their existence, that what applies to the one is almost sure not to apply to the other. Domestic animals are abnormal, irregular, artificial; they are subject to varieties which never occur and never can occur in a state of nature: their very existence depends altogether on human care; so far are many of them removed from that just proportion of faculties, that true balance of organization, by means of which alone an animal left to its own resources can preserve its existence and continue its race.

The hypothesis of Lamarck – that progressive changes in species have been produced by the attempts of animals to increase the development of their own organs, and thus modify their structure and habits – has been repeatedly and easily refuted by all writers on the subject of varieties and species, and it seems to have been considered that when this was done the whole question has been finally settled; but the view here developed renders such an hypothesis quite unnecessary, by showing that similar results must be produced by the action of principles constantly at work in nature. The powerful retractile talons of the falcon- and the cat-tribes have not been produced or increased by the volition of those animals; but among the different varieties which occurred in the earlier and less highly organized forms of these groups, *those always survived longest which had the greatest facilities for seizing their prey.* Neither did the giraffe acquire its long neck by desiring to reach the foliage of the more lofty shrubs, and constantly stretching its neck for the purpose, but because any varieties which occurred among its antitypes with a longer neck than usual *at once secured a fresh range of pasture over the same ground as their shorter-necked companions, and on the first scarcity of food were thereby enabled to outlive them.* Even the peculiar colours of many animals, especially insects, so closely resembling the soil or the leaves or the trunks on which they habitually reside, are explained on the same principle; for though in the course of ages varieties of many tints may have occurred, *yet those races having colours best adapted to concealment from their enemies would inevitably survive the longest.* We have also here an acting cause to account for that balance so often observed in nature, – a deficiency in one set of organs always being compensated by an increased development of some others – powerful wings accompanying weak feet, or great velocity making up for the absence of defensive weapons; for it has been shown that all varieties in which an unbalanced deficiency occurred could not long continue their existence. The action of this principle is exactly like that of the centrifugal governor of the steam engine, which checks and corrects any irregularities almost before they become evident; and in like manner no unbalanced deficiency in the animal kingdom can ever reach any conspicuous magnitude, because it would make itself felt at the very first step, by rendering existence difficult and extinction almost sure soon to follow. An origin such as is here advocated will also agree with the peculiar character of the modifications of form and structure which obtain in organized beings – the many lines of divergence from a central type, the increasing efficiency and power of a particular organ through a succession of allied species, and the remarkable persistence of unimportant parts such as colour, texture of plumage and hair, form of horns or crests, through a series of species differing considerably in more essential characters. It also furnishes us with a reason for that "more specialized structure" which Professor Owen states to be a characteristic of recent compared with extinct forms, and which would evidently be the result of the progressive modification of any organ applied to a special purpose in the animal economy.

We believe we have now shown that there is a tendency in nature to the continued progression of certain classes of *varieties* further and further from the original type – a progression to which there appears no reason to assign any definite limits – and that the same principle which produces this result in a state of nature will also explain why domestic varieties have a tendency to revert to the original type. This progression, by minute steps, in various directions, but always checked and balanced by the necessary conditions, subject to which alone existence can be preserved, may, it is believed, be followed out so as to agree with all the phenomena presented by organized beings, their extinction and succession in past ages, and all the extraordinary modifications of form, instinct, and habits which they exhibit.

•

It has sometimes been said that the success of the Origin *proved "that the subject was in the air," or "that men's minds were prepared for it." I do not think that this is strictly true, ... What I believe was strictly true is that innumerable well-observed facts were stored in the minds of naturalists, ready to take their proper places as soon as any theory which would receive them was sufficiently explained.*

Charles Darwin, Autobiography

1859

Charles Darwin
Recapitulation and Conclusion

[Charles Darwin, "Recapitulation and Conclusion," in *On the Origin of Species.* London: John Murray, 1859, Ch. 14, pp. 459-490]

AS THIS WHOLE VOLUME is one long argument, it may be convenient to the reader to have the leading facts and inferences briefly recapitulated.

That many and grave objections may be advanced against the theory of descent with modification through natural selection, I do not deny. I have endeavoured to give to them their full force. Nothing at first can appear more difficult to believe than that the more complex organs and instincts should have been perfected, not by means superior to, though analogous with, human reason, but by the accumulation of innumerable slight variations, each good for the individual possessor. Nevertheless, this difficulty, though appearing to our imagination insuperably great, cannot be considered real if we admit the following propositions, namely, – that gradations in the perfection of any organ or instinct, which we may consider, either do now exist or could have existed, each good of its kind, – that all organs and instincts are, in ever so slight a degree, variable, – and, lastly, that there is a struggle for existence leading to the preservation of each profitable deviation of structure or instinct. The truth of these propositions cannot, I think, be disputed.

It is, no doubt, extremely difficult even to conjecture by what gradations many structures have been perfected, more especially amongst broken and failing groups of organic beings; but we see so many strange gradations in nature, as is proclaimed by the canon, "Natura non facit saltum," that we ought to be extremely cautious in saying that any organ or instinct, or any whole being, could not have arrived at its present state by many graduated steps. There are, it must be admitted, cases of special difficulty on the theory of natural selection; and one of the most curious of these is the existence of two or three defined castes of workers or sterile females in the same community of ants; but I have attempted to show how this difficulty can be mastered.

With respect to the almost universal sterility of species when first crossed, which forms so remarkable a contrast with the almost universal fertility of varieties when crossed, I must refer the reader to the recapitulation of the facts given at the end of the eighth chapter, which seem to me conclusively to show that this sterility is no more a special endowment than is the incapacity of two trees to be grafted together; but that it is incidental on constitutional differences in the reproductive systems of the intercrossed species. We see the truth of this conclusion in the vast difference in the result, when the same two species are crossed reciprocally; that is, when one species is first used as the father and then as the mother.

The fertility of varieties when intercrossed and of their mongrel offspring cannot be considered as universal; nor is their very general fertility surprising when we remember that it is not likely that either their constitutions or their reproductive systems should have been profoundly modified. Moreover, most of the varieties which have been experimentised on have been produced under domestication; and as domestication apparently tends to eliminate sterility, we ought not to expect it also to produce sterility.

The sterility of hybrids is a very different case from that of first crosses, for their reproductive organs are more or less functionally impotent; whereas in first crosses the organs on both sides are in a perfect condition. As we continually see that organisms of all kinds are rendered in some degree sterile from their constitutions having been disturbed by slightly different and new conditions of life, we need not feel surprise at hybrids being in some degree sterile, for their constitutions can hardly fail to have been disturbed from being compounded of two distinct organisations. This parallelism is supported by another parallel, but directly opposite, class of facts; namely, that the vigour and fertility of all organic beings are increased by slight changes in their conditions of life, and that the offspring of slightly modified forms or varieties acquire from being crossed increased vigour and fertility. So that, on the one hand, considerable changes in the conditions of life and crosses between greatly modified forms, lessen fertility; and on the other hand, lesser changes in the conditions of life and crosses between less modified forms, increase fertility.

Turning to geographical distribution, the difficulties encountered on the theory of descent with modification are grave enough. All the individuals of the same species, and all the species of the same genus, or even higher group, must have descended from common parents; and therefore, in however distant and isolated parts of the world they are now found, they must in the course of successive generations have passed from some one part to the others. We are often wholly unable even to conjecture how this could have been effected. Yet, as we have reason to believe that some species have retained the same specific form for very long periods, enormously long as measured by years, too much stress ought not to be laid on the occasional wide diffusion of the same species; for during very long periods of time there will always be a good chance for wide migration by many means. A broken or interrupted range may often be accounted for by the extinction of the species in the intermediate regions. It cannot be denied that we are as yet very ignorant of the full extent of the various climatal and geographical changes which have affected the earth during modern periods; and such changes will obviously have greatly facilitated migration. As an example, I have attempted to show how potent has been the influence of the Glacial period on the distribution both of the same and of representative species throughout the world. We are as yet profoundly ignorant of the many occasional means of transport. With respect to distinct species of the same genus inhabiting very distant and isolated regions, as the process of modification has necessarily been slow, all the means of migration will have been possible during a very long period; and consequently the difficulty of the wide diffusion of species of the same genus is in some degree lessened.

As on the theory of natural selection an interminable number of intermediate forms must have existed, linking together all the species in each group by gradations as fine as our present varieties, it may be asked, Why do we not see these linking forms all around us? Why are not all organic beings blended together in an inextricable chaos? With respect to existing forms, we should remember that we have no right to expect (excepting in rare cases) to discover *directly* connecting links between them, but only between each and some extinct and supplanted form. Even on a wide area, which has during a long period remained continuous, and of which the climate and other conditions of life change insensibly in going from a district occupied by one species into another district occupied by a closely allied species, we have no just right to expect often to find intermediate varieties in the intermediate zone. For we have reason to believe that only a few species are undergoing change at any one period; and all changes are slowly effected. I have also shown that the intermediate varieties which will at first probably exist in the intermediate zones, will be liable to be supplanted by the allied forms on either hand; and the latter, from existing in greater numbers, will generally be modified and improved at a quicker rate than the intermediate varieties, which exist in lesser numbers; so that the intermediate varieties will, in the long run, be supplanted and exterminated.

On this doctrine of the extermination of an infinitude of connecting links, between the living and extinct inhabitants of the world, and at each successive period between the extinct and still older species, why is not every geological formation charged with such links? Why does not every collection of fossil remains afford plain evidence of the gradation and mutation of the forms of life? We meet with no such evidence, and this is the most obvious and forcible of the many objections which may be urged against my theory. Why, again, do whole groups of allied species appear, though certainly they often falsely appear, to have come in suddenly on the several geological stages? Why do we not find great piles of strata beneath the Silurian system, stored with the remains of the progenitors of the Silurian groups of fossils? For certainly on my theory such strata must somewhere have been deposited at these ancient and utterly unknown epochs in the world's history.

I can answer these questions and grave objections only on the supposition that the geological record is far more imperfect than most geologists believe. It cannot be objected that there has not been time sufficient for any amount of organic change; for the lapse of time has been so great as to be utterly inappreciable by the human intellect. The number of specimens in all our museums is absolutely as nothing compared with the countless generations of countless species which certainly have existed. We should not be able to recognise a species as the parent of any one or more species if we were to examine them ever so closely, unless we likewise possessed many of the intermediate links between their past or parent and present states; and these many links we could hardly ever expect to discover, owing to the imperfection of the geological record. Numerous existing doubtful forms could be named which are probably varieties; but who will pretend that in future ages so many fossil links will be discovered, that naturalists will be able to decide, on the common view, whether or not these doubtful forms are varieties? As long as most of the links between any two species are unknown, if any one link or intermediate variety be discovered, it will simply be classed as another and distinct species. Only a small portion of the world has been geologically explored. Only organic beings of certain classes can be preserved in a fossil condition, at least in any great number. Widely ranging species vary most, and varieties are often at first local, – both causes rendering the discovery of intermediate links less likely. Local varieties will not spread into other and distant regions until they are considerably modified and improved; and when they do spread, if discovered in a geological formation, they will appear as if suddenly created there, and will be simply classed as new species. Most formations have been intermittent in their accumulation; and their duration, I am inclined to believe, has been shorter than the average duration of specific forms. Successive formations are separated from each other by enormous blank intervals of time; for fossiliferous formations, thick enough to resist future degradation, can be accumulated only where much sediment is deposited on the subsiding bed of the sea. During the alternate periods of elevation and of stationary level the record will be blank. During these latter periods there will probably be more variability in the forms of life; during periods of subsidence, more extinction.

With respect to the absence of fossiliferous formations beneath the lowest Silurian strata, I can only recur to the hypothesis given in the ninth chapter. That the geological record is imperfect all will admit; but that it is imperfect to the degree which I require, few will be inclined to admit. If we look to long enough intervals of time, geology plainly declares that all species have changed; and they have changed in the manner which my theory requires, for they have changed slowly and in a graduated manner. We clearly see this in the fossil remains from consecutive formations invariably being much more closely related to each other, than are the fossils from formations distant from each other in time.

Such is the sum of the several chief objections and difficulties which may justly be urged against my theory; And I have now briefly recapitulated the answers and explanations which can be given to them. I have felt these difficulties far too heavily during many years to doubt their weight. But it deserves especial notice that the more important objections relate to questions on which we are confessedly ignorant;

nor do we know how ignorant we are. We do not know all the possible transitional gradations between the simplest and the most perfect organs; it cannot be pretended that we know all the varied means of Distribution during the long lapse of years, or that we know how imperfect the Geological Record is. Grave as these several difficulties are, in my judgment they do not overthrow the theory of descent with modification.

Now let us turn to the other side of the argument. Under domestication we see much variability. This seems to be mainly due to the reproductive system being eminently susceptible to changes in the conditions of life; so that this system, when not rendered impotent, fails to reproduce offspring exactly like the parent-form. Variability is governed by many complex laws, – by correlation of growth, by use and disuse, and by the direct action of the physical conditions of life. There is much difficulty in ascertaining how much modification our domestic productions have undergone; but we may safely infer that the amount has been large, and that modifications can be inherited for long periods. As long as the conditions of life remain the same, we have reason to believe that a modification, which has already been inherited for many generations, may continue to be inherited for an almost infinite number of generations. On the other hand we have evidence that variability, when it has once come into play, does not wholly cease; for new varieties are still occasionally produced by our most anciently domesticated productions.

Man does not actually produce variability; he only unintentionally exposes organic beings to new conditions of life, and then nature acts on the organisation, and causes variability. But man can and does select the variations given to him by nature, and thus accumulate them in any desired manner. He thus adapts animals and plants for his own benefit or pleasure. He may do this methodically, or he may do it unconsciously by preserving the individuals most useful to him at the time, without any thought of altering the breed. It is certain that he can largely influence the character of a breed by selecting, in each successive generation, individual differences so slight as to be quite inappreciable by an uneducated eye. This process of selection has been the great agency in the production of the most distinct and useful domestic breeds. That many of the breeds produced by man have to a large extent the character of natural species, is shown by the inextricable doubts whether very many of them are varieties or aboriginal species.

There is no obvious reason why the principles which have acted so efficiently under domestication should not have acted under nature. In the preservation of favoured individuals and races, during the constantly-recurrent Struggle for Existence, we see the most powerful and ever-acting means of selection. The struggle for existence inevitably follows from the high geometrical ratio of increase which is common to all organic beings. This high rate of increase is proved by calculation, by the effects of a succession of peculiar seasons, and by the results of naturalisation, as explained in the third chapter. More individuals are born than can possibly survive. A grain in the balance will determine which individual shall live and which shall die, – which variety or species shall increase in number, and which shall decrease, or finally become extinct. As the individuals of the same species come in all respects into the closest competition with each other, the struggle will generally be most severe between them; it will be almost equally severe between the varieties of the same species, and next in severity between the species of the same genus. But the struggle will often be very severe between beings most remote in the scale of nature. The slightest advantage in one being, at any age or during any season, over those with which it comes into competition, or better adaptation in however slight a degree to the surrounding physical conditions, will turn the balance.

With animals having separated sexes there will in most cases be a struggle between the males for possession of the females. The most vigorous individuals, or those which have most successfully struggled with their conditions of life, will generally leave most progeny. But success will often depend on having special weapons or means of defence, or on the charms of the males; and the slightest advantage will lead to victory.

As geology plainly proclaims that each land has undergone great physical changes, we might have expected that organic beings would have varied under nature, in the same way as they generally have varied under the changed conditions of domestication. And if there be any variability under nature, it would be an unaccountable fact if natural selection had not come into play. It has often been asserted, but the assertion is quite incapable of proof, that the amount of variation under nature is a strictly limited quantity. Man, though acting on external characters alone and often capriciously, can produce within a short period a great result by adding up mere individual differences in his domestic productions; and every one admits that there are at least individual differences in species under nature. But, besides such differences, all naturalists have admitted the existence of varieties, which they think sufficiently distinct to be worthy of record in systematic works. No one can draw any clear distinction between individual differences and slight varieties; or between more plainly marked varieties and sub-species, and species. Let it be observed how naturalists differ in the rank which they assign to the many representative forms in Europe and North America.

If then we have under nature variability and a powerful agent always ready to act and select, why should we doubt that variations in any way useful to beings, under their excessively complex relations of life, would be preserved, accumulated, and inherited? Why, if man can by patience select variations most useful to himself, should nature fail in selecting variations useful, under changing conditions of life, to her living products? What limit can be put to this power, acting during long ages and rigidly scrutinising the whole constitution, structure, and habits of each creature, – favouring the good and rejecting the bad? I can see no limit to this power, in slowly and beautifully adapting each form to the most complex relations of life. The theory of natural selection, even if we looked no further than this, seems to me to be in itself probable. I have already recapitulated, as fairly as I could, the opposed difficulties and objections: now let us turn to the special facts and arguments in favour of the theory.

On the view that species are only strongly marked and permanent varieties, and that each species first existed as a variety, we can see why it is that no line of demarcation can be drawn between species, commonly supposed to have been produced by special acts of creation, and varieties which are acknowledged to have been produced by secondary laws. On this same view we can understand how it is that in each region where many species of a genus have been produced, and where they now flourish, these same species should present many varieties; for where the manufactory of species has been active, we might expect, as a general rule, to find it still in action; and this is the case if varieties be incipient species. Moreover, the species of the larger genera, which afford the greater number of varieties or incipient species, retain to a certain degree the character of varieties; for they differ from each other by a less amount of difference than do the species of smaller genera. The closely allied species also of the larger genera apparently have restricted ranges, and they are clustered in little groups round other species – in which respects they resemble varieties. These are strange relations on the view of each species having been independently created, but are intelligible if all species first existed as varieties.

As each species tends by its geometrical ratio of reproduction to increase inordinately in number; and as the modified descendants of each species will be enabled to increase by so much the more as they become more diversified in habits and structure, so as to be enabled to seize on many and widely different places in the economy of nature, there will be a constant tendency in natural selection to preserve the most divergent offspring of any one species. Hence during a long-continued course of modification, the slight differences, characteristic of varieties of the same species, tend to be augmented into the greater differences characteristic of species of the same genus. New and improved varieties will inevitably supplant and exterminate the older, less improved and intermediate varieties; and thus species are rendered to a large extent defined and distinct objects. Dominant species belonging to the larger groups tend to give birth to new and dominant forms; so that each large group tends to become still larger, and at the same time more divergent in character. But as all groups cannot thus succeed in increasing in size, for the world would not hold them, the more dominant groups beat the less dominant. This tendency in the large groups to go on increasing in size and diverging in character, together with the almost inevitable contingency of much extinction, explains the arrangement of all the forms of life, in groups subordinate to groups, all within a few great classes, which we now see everywhere around us, and which has prevailed throughout all time. This grand fact of the grouping of all organic beings seems to me utterly inexplicable on the theory of creation.

As natural selection acts solely by accumulating slight, successive, favourable variations, it can produce no great or sudden modification; it can act only by very short and slow steps. Hence the canon of "Natura non facit saltum," which every fresh addition to our knowledge tends to make more strictly correct, is on this theory simply intelligible. We can plainly see why nature is prodigal in variety, though niggard in innovation. But why this should be a law of nature if each species has been independently created, no man can explain.

Many other facts are, as it seems to me, explicable on this theory. How strange it is that a bird, under the form of woodpecker, should have been created to prey on insects on the ground; that upland geese, which never or rarely swim, should have been created with webbed feet; that a thrush should have been created to dive and feed on sub-aquatic insects; and that a petrel should have been created with habits and structure fitting it for the life of an auk or grebe! and so on in endless other cases. But on the view of each species constantly trying to increase in number, with natural selection always ready to adapt the slowly varying descendants of each to any unoccupied or ill-occupied place in nature, these facts cease to be strange, or perhaps might even have been anticipated.

As natural selection acts by competition, it adapts the inhabitants of each country only in relation to the degree of perfection of their associates; so that we need feel no surprise at the inhabitants of any one country, although on the ordinary view supposed to have been specially created and adapted for that country, being beaten and supplanted by the naturalised productions from another land. Nor ought we to marvel if all the contrivances in nature be not, as far as we can judge, absolutely perfect; and if some of them be abhorrent to our ideas of fitness. We need not marvel at the sting of the bee causing the bee's own death; at drones being produced in such vast numbers for one single act, and being then slaughtered by their sterile sisters; at the astonishing waste of pollen by our fir trees; at the instinctive hatred of the queen bee for her own fertile daughters; at ichneumonidae feeding within the live bodies of caterpillars; and at other such cases. The wonder indeed is, on the theory of natural selection, that more cases of the want of absolute perfection have not been observed.

The complex and little known laws governing variation are the same, as far as we can see, with the laws which have governed the production of so-called specific forms. In both cases physical conditions seem to have produced but little direct effect; yet when varieties enter any zone, they occasionally assume some of the characters of the species proper to that zone. In both varieties and species, use and disuse seem to have produced some effect; for it is difficult to resist this conclusion when we look, for instance, at the logger-headed duck, which has wings incapable of flight, in nearly the same condition as in the domestic duck; or when we look at the burrowing tucutucu, which is occasionally blind, and then at certain moles, which are habitually blind and have their eyes covered with skin; or when we look at the blind animals inhabiting the dark caves of America and Europe. In both varieties and species correlation of growth seems to have played a most important part, so that when one part has been modified other parts are necessarily modified. In both varieties and species reversions to long-lost characters occur. How inexplicable on the theory of creation is the occasional appearance of stripes on the shoulder and legs of the several species of the horse-genus and in their hybrids! How simply is this fact explained if we believe that these species have descended from a striped progenitor, in the same manner as the several domestic breeds of pigeon have descended from the blue and barred rock-pigeon!

On the ordinary view of each species having been independently created, why should the specific characters, or those by which the species of the same genus differ from each other, be more variable than the generic characters in which they all agree? Why, for instance, should the colour of a flower be more likely to vary in any one species of a genus, if the other species, supposed to have been created independently, have differently coloured flowers, than if all the species of the genus have the same coloured flowers? If species are only well-marked varieties, of which the characters have become in a high degree permanent, we can understand this fact; for they have already varied since they branched off from a common progenitor in certain characters, by which they have come to be specifically distinct from each other; and therefore these same characters would be more likely still to be variable than the generic characters which have been inherited without change for an enormous period. It is inexplicable on the theory of creation why a part developed in a very unusual manner in any one species of a genus, and therefore, as we may naturally infer, of great importance to the species, should be eminently liable to variation; but, on my view, this part has undergone, since the several species branched off from a common progenitor, an unusual amount of variability and modification, and therefore we might expect this part generally to be still variable. But a part may be developed in the most unusual manner, like the wing of a bat, and yet not be more variable than any other structure, if the part be common to many subordinate forms, that is, if it has been inherited for a very long period; for in this case it will have been rendered constant by long-continued natural selection.

Glancing at instincts, marvellous as some are, they offer no greater difficulty than does corporeal structure on the theory of the natural selection of successive, slight, but profitable modifications. We can thus understand why nature moves by graduated steps in endowing different animals of the same class with their several instincts. I have attempted to show how much light the principle of gradation throws on the admirable architectural powers of the hive-bee. Habit no doubt sometimes comes into play in modifying instincts; but it certainly is not indispensable, as we see, in the case of neuter insects, which leave no progeny to inherit the effects of long-continued habit. On the view of all the species of the same genus having descended from a common parent, and having inherited much in common, we can understand how it is that allied species, when placed under considerably different conditions of life, yet should follow nearly the same instincts; why the thrush of South America, for instance, lines her nest with mud like our British species. On the view of instincts having been slowly acquired through natural selection we need not marvel at some instincts being apparently not perfect and liable to mistakes, and at many instincts causing other animals to suffer.

If species be only well-marked and permanent varieties, we can at once see why their crossed offspring should follow the same complex laws in their degrees and kinds of resemblance to their parents, – in being absorbed into each other by successive crosses, and in other such points, – as do the crossed offspring of acknowledged varieties. On the other hand, these would be strange facts if species have been independently created, and varieties have been produced by secondary laws.

If we admit that the geological record is imperfect in an extreme degree, then such facts as the record gives, support the theory of descent with modification. New species have come on the stage slowly and at successive intervals; and the amount of change, after equal intervals of time, is widely different in different groups. The extinction of species and of whole groups of species, which has played so conspicuous a part in the history of the organic world, almost inevitably follows on the principle of natural selection; for old forms will be supplanted by new and improved forms. Neither single species nor groups of species reappear when the chain of ordinary generation has once been broken. The gradual diffusion of dominant forms, with the slow modification of their descendants, causes the forms of life, after long intervals of time, to appear as if they had changed simultaneously throughout the world. The fact of the fossil remains of each formation being in some degree intermediate in character between the fossils in the formations above and below, is simply explained by their intermediate position in the chain of descent. The grand fact that all extinct organic beings belong to the same system with recent beings, falling either into the same or into intermediate groups, follows from the living and the extinct being the offspring of common parents. As the groups which have descended from an ancient progenitor have generally diverged in character, the progenitor with its early descendants will often be intermediate in character in comparison with its later descendants; and thus we can see why the more ancient a fossil is, the oftener it stands in some degree intermediate between existing and allied groups. Recent forms are generally looked at as being, in some vague sense, higher than ancient and extinct forms; and they are in so far higher as the later and more improved forms have conquered the older and less improved organic beings in the struggle for life. Lastly, the law of the long endurance of allied forms on the same continent, – of marsupials in Australia, of edentata in America, and other such cases, – is intelligible, for within a confined country, the recent and the extinct will naturally be allied by descent.

Looking to geographical distribution, if we admit that there has been during the long course of ages much migration from one part of the world to another, owing to former climatal and geographical changes and to the many occasional and unknown means of dispersal, then we can understand, on the theory of descent with modification, most of the great leading facts in Distribution. We can see why there should be so striking a parallelism in the distribution of organic beings throughout space, and in their geological succession throughout time; for in both cases the beings have been connected by the bond of ordinary generation, and the means of modification have been the same. We see the full meaning of the wonderful fact, which must have struck every traveller, namely, that on the same continent, under the most diverse conditions, under heat and cold, on mountain and lowland, on deserts and marshes, most of the inhabitants within each great class are plainly related; for they will generally be descendants of the same progenitors and early colonists. On this same principle of former migration, combined in most cases with modification, we can understand, by the aid of the Glacial period, the identity of some few plants, and the close alliance of many others, on the most distant mountains, under the most different climates; and likewise the close alliance of some of the inhabitants of the sea in the northern and southern temperate zones, though separated by the whole intertropical ocean. Although two areas may present the same physical conditions of life, we need feel no surprise at their inhabitants being widely different, if they have been for a long period completely separated from each other; for as the relation of organism to organism is the most important of all relations, and as the two areas will have received colonists from some third source or from each other, at various periods and in different proportions, the course of modification in the two areas will inevitably be different.

On this view of migration, with subsequent modification, we can see why oceanic islands should be inhabited by few species, but of these, that many should be peculiar. We can clearly see why those animals which cannot cross wide spaces of ocean, as frogs and terrestrial mammals, should not inhabit oceanic islands; and why, on the other hand, new and peculiar species of bats, which can traverse the ocean, should so often be found on islands far distant from any continent. Such facts as the presence of peculiar species of bats, and the absence of all other mammals, on oceanic islands, are utterly inexplicable on the theory of independent acts of creation.

The existence of closely allied or representative species in any two areas, implies, on the theory of descent with modification, that the same parents formerly inhabited both areas; and we almost invariably find that wherever many closely allied species inhabit two areas, some identical species common to both still exist. Wherever many closely allied yet distinct species occur, many doubtful forms and varieties of the same species likewise occur. It is a rule of high generality that the inhabitants of each area are related to the inhabitants of the nearest source whence immigrants might have been derived. We see this in nearly all the plants and animals of the Galapagos archipelago, of Juan Fernandez, and of the other American islands being related in the most striking manner to the plants and animals of the neighbouring American mainland; and those of the Cape de Verde archipelago and other African islands to the African mainland. It must be admitted that these facts receive no explanation on the theory of creation.

The fact, as we have seen, that all past and present organic beings constitute one grand natural system, with group subordinate to group, and with extinct groups often falling in between recent groups, is intelligible on the theory of natural selection with its contingencies of extinction and divergence of character. On these same principles we see how it is, that the mutual affinities of the species and genera within each class are so complex and circuitous. We see why certain characters are far more serviceable than others for classification; – why adaptive characters, though of paramount importance to the being, are of hardly any importance in classification; why characters derived from rudimentary parts, though of no service to the being, are often of high classificatory value; and why embryological characters are the most valuable of all. The real affinities of all organic beings are due to inheritance or community of descent. The natural system is a genealogical arrangement, in which we have to discover the lines of descent by the most permanent characters, however slight their vital importance may be.

The framework of bones being the same in the hand of a man, wing of a bat, fin of the porpoise, and leg of the horse, – the same number of vertebrae forming the neck of the giraffe and of the elephant, – and innumerable other such facts, at once explain themselves on the theory of descent with slow and slight successive modifications. The similarity of pattern in the wing and leg of a bat, though used for such different purpose, – in the jaws and legs of a crab, – in the petals, stamens, and pistils of a flower, is likewise intelligible on the view of the gradual modification of parts or organs, which were alike in the early progenitor of each class. On the principle of successive variations not always supervening at an early age, and being inherited at a corresponding not early period of life, we can clearly see why the embryos of mammals, birds, reptiles, and fishes should be so closely alike, and should be so unlike the adult forms. We may cease marvelling at the embryo of an air-breathing mammal or bird having branchial slits and arteries running in loops, like those in a fish which has to breathe the air dissolved in water, by the aid of well-developed branchiae.

Disuse, aided sometimes by natural selection, will often tend to reduce an organ, when it has become useless by changed habits or under changed conditions of life; and we can clearly understand on this view the meaning of rudimentary organs. But disuse and selection will generally act on each creature, when it has come to maturity and has to play its full part in the struggle for existence, and will thus have little power of acting on an organ during early life; hence the organ will not be much reduced or rendered rudimentary at this early age. The calf, for instance, has inherited teeth, which never cut through the gums of the upper jaw, from an early progenitor having well-developed teeth; and we may believe, that the teeth in the mature animal were reduced, during successive generations by disuse or by the tongue and palate having been fitted by natural selection to browse without their aid; whereas in the calf, the teeth have been left untouched by selection or disuse, and on the principle of inheritance at corresponding ages have been inherited from a remote period to the present day. On the view of each organic being and each separate organ having been specially created, how utterly inexplicable it is that parts, like the teeth in the embryonic calf or like the shrivelled wings under the soldered wing-covers of some beetles, should thus so frequently bear the plain stamp of utility! Nature may be said to have taken pains to reveal, by rudimentary organs and by homologous structures, her scheme of modification, which it seems that we wilfully will not understand.

I have now recapitulated the chief facts and considerations which have thoroughly convinced me that species have changed, and are still slowly changing by the preservation and accumulation of successive slight favourable variations. Why, it may be asked, have all the most eminent living naturalists and geologists rejected this view of the mutability of species? It cannot be asserted that organic beings in a state of nature are subject to no variation; it cannot be proved that the amount of variation in the course of long ages is a limited quantity; no clear distinction has been, or can be, drawn between species and well-marked varieties. It cannot be maintained that species when intercrossed are invariably sterile, and varieties invariably fertile; or that sterility is a special endowment and sign of creation. The belief that species were immutable productions was almost unavoidable as long as the history of the world was thought to be of short duration; and now that we have acquired some idea of the lapse of time, we are too apt to assume, without proof, that the geological record is so perfect that it would have afforded us plain evidence of the mutation of species, if they had undergone mutation.

But the chief cause of our natural unwillingness to admit that one species has given birth to other and distinct species, is that we are always slow in admitting any great change of which we do not see the intermediate steps. The difficulty is the same as that felt by so many geologists, when Lyell first insisted that long lines of inland cliffs had been formed, and great valleys excavated, by the slow action of the coast-waves. The mind cannot possibly grasp the full meaning of the term of a hundred million years; it cannot add up and perceive the full effects of many slight variations, accumulated during an almost infinite number of generations.

Although I am fully convinced of the truth of the views given in this volume under the form of an abstract, I by no means expect to convince experienced naturalists whose minds are stocked with a multitude of facts all viewed, during a long course of years, from a point of view directly opposite to mine. It is so easy to hide our ignorance under such expressions as the "plan of creation," "unity of design," &c., and to think that we give an explanation when we only restate a fact. Any one whose disposition leads him to attach more weight to unexplained difficulties than to the explanation of a certain number of facts will certainly reject my theory. A few naturalists, endowed with much flexibility of mind, and who have already begun to doubt on the immutability of species, may be influenced by this volume; but I look with confidence to the future, to young and rising naturalists, who will be able to view both sides of the question with impartiality. Whoever is led to believe that species are mutable will do good service by conscientiously expressing his conviction; for only thus can the load of prejudice by which this subject is overwhelmed be removed.

Several eminent naturalists have of late published their belief that a multitude of reputed species in each genus are not real species; but that other species are real, that is, have been independently created. This seems to me a strange conclusion to arrive at. They admit that a multitude of forms, which till lately they themselves thought were special creations, and which are still thus looked at by the majority of naturalists, and which consequently have every external characteristic feature of true species, – they admit that these have been produced by variation, but they refuse to extend the same view to other and very slightly different forms. Nevertheless they do not pretend that they can define, or even conjecture, which are the created forms of life, and which are those produced by secondary laws. They admit variation as a *vera causa* in one case, they arbitrarily reject it in another, without assigning any distinction in the two cases. The day will come when this will be given as a curious illustration of the blindness of preconceived opinion. These authors seem no more startled at a miraculous act of creation than at an ordinary birth. But do they really believe that at innumerable periods in the earth's history certain elemental atoms have been commanded suddenly to flash into living tissues? Do they believe that at each supposed act of creation one individual or many were produced? Were all the infinitely numerous kinds of animals and plants created as eggs or seed, or as full grown? and in the case of mammals, were they created bearing the false marks of nourishment from the mother's womb? Although naturalists very properly demand a full explanation of every difficulty from those who believe in the mutability of species, on their own side they ignore the whole subject of the first appearance of species in what they consider reverent silence.

It may be asked how far I extend the doctrine of the modification of species. The question is difficult to answer, because the more distinct the forms are which we may consider, by so much the arguments fall away in force. But some arguments of the greatest weight extend very far. All the members of whole classes can be connected together by chains of affinities, and all can be classified on the same principle, in groups subordinate to groups. Fossil remains sometimes tend to fill up very wide intervals between existing orders. Organs in a rudimentary condition plainly show that an early progenitor had the organ in a fully developed state; and this in some instances necessarily implies an enormous amount of modification in the descendants. Throughout whole classes various structures are formed on the same pattern, and at an embryonic age the species closely resemble each other. Therefore I cannot doubt that the theory of descent with modification embraces all the members of the same class. I believe that animals have descended from at most only four or five progenitors, and plants from an equal or lesser number.

Analogy would lead me one step further, namely, to the belief that all animals and plants have descended from some one prototype. But analogy may be a deceitful guide. Nevertheless all living things have much in common, in their chemical composition, their germinal vesicles, their cellular structure, and their laws of growth and reproduction. We see this even in so trifling a circumstance as that the same poison often similarly affects plants and animals; or that the poison secreted by the gall-fly produces monstrous growths on the wild rose or oak-tree. Therefore I should infer from analogy that probably all the organic beings which have ever lived on this earth have descended from some one primordial form, into which life was first breathed.

When the views entertained in this volume on the origin of species, or when analogous views are generally admitted, we can dimly foresee that there will be a considerable revolution in natural history. Systematists will be able to pursue their labours as at present; but they will not be incessantly haunted by the shadowy doubt whether this or that form be in essence a species. This I feel sure, and I speak after experience, will be no slight relief. The endless disputes whether or not some fifty species of British brambles are true species will cease. Systematists will have only to decide (not that this will be easy) whether any form be sufficiently constant and distinct from other forms, to be capable of definition; and if definable, whether the differences be sufficiently important to deserve a specific name. This latter point will become a far more essential consideration than it is at present; for differences, however slight, between any two forms, if not blended by intermediate gradations, are looked at by most naturalists as sufficient to raise both forms to the rank of species. Hereafter we shall be compelled to acknowledge that the only distinction between species and well-marked varieties is, that the latter are known, or believed, to be connected at the present day by intermediate gradations, whereas species were formerly thus connected. Hence, without quite rejecting the consideration of the present existence of intermediate gradations between any two forms, we shall be led to weigh more carefully and to value higher the actual amount of difference between them. It is quite possible that forms now generally acknowledged to be merely varieties may hereafter be thought worthy of specific names, as with the primrose and cowslip; and in this case scientific and common language will come into accordance. In short, we shall have to treat species in the same manner as those naturalists treat genera, who admit that genera are merely artificial combinations made for convenience. This may not be a cheering prospect; but we shall at least be freed from the vain search for the undiscovered and undiscoverable essence of the term species.

The other and more general departments of natural history will rise greatly in interest. The terms used by naturalists of affinity, relationship, community of type, paternity, morphology, adaptive characters, rudimentary and aborted organs, &c., will cease to be metaphorical, and will have a plain signification. When we no longer look at an organic being as a savage looks at a ship, as at something wholly beyond his comprehension; when we regard every production of nature as one which has had a history; when we contemplate every complex structure and instinct as the summing up of many contrivances, each useful to the possessor, nearly in the same way as when we look at any great mechanical invention as the summing up of the labour, the experience, the reason, and even the blunders of numerous workmen; when we thus view each organic being, how far more interesting, I speak from experience, will the study of natural history become!

A grand and almost untrodden field of inquiry will be opened, on the causes and laws of variation, on correlation of growth, on the effects of use and disuse, on the direct action of external conditions, and so forth. The study of domestic productions will rise immensely in value. A new variety raised by man will be a far more important and interesting subject for study than one more species added to the infinitude of already recorded species. Our classifications will come to be, as far as they can be so made, genealogies; and will then truly give what may be called the plan of creation. The rules for classifying will no doubt become simpler when we have a definite object in view. We possess no pedigrees or armorial bearings; and we have to discover and trace the many diverging lines of descent in our natural genealogies, by characters of any kind which have long been inherited. Rudimentary organs will speak infallibly with respect to the nature of long-lost structures. Species and groups of species, which are called aberrant, and which may fancifully be called living fossils, will aid us in forming a picture of the ancient forms of life. Embryology will reveal to us the structure, in some degree obscured, of the prototypes of each great class.

When we can feel assured that all the individuals of the same species, and all the closely allied species of most genera, have within a not very remote period descended from one parent, and have migrated from some one birthplace; and when we better know the many means of migration, then, by the light which geology now throws, and will continue to throw, on former changes of climate and of the level of the land, we shall surely be enabled to trace in an admirable manner the former migrations of the inhabitants of the whole world. Even at present, by comparing the differences of the inhabitants of the sea on the opposite sides of a continent, and the nature of the various inhabitants of that continent in relation to their apparent means of immigration, some light can be thrown on ancient geography.

The noble science of Geology loses glory from the extreme imperfection of the record. The crust of the earth with its embedded remains must not be looked at as a well-filled museum, but as a poor collection made at hazard and at rare intervals. The accumulation of each great fossiliferous formation will be recognised as having depended on an unusual concurrence of circumstances, and the blank intervals between the successive stages as having been of vast duration. But we shall be able to gauge with some security the duration of these intervals by a comparison of the preceding and succeeding organic forms. We must be cautious in attempting to correlate as strictly contemporaneous two formations, which include few identical species, by the general succession of their forms of life. As species are produced and exterminated by slowly acting and still existing causes, and not by miraculous acts of creation and by catastrophes; and as the most important of all causes of organic change is one which is almost independent of altered and perhaps suddenly altered physical conditions, namely, the mutual relation of organism to organism, – the improvement of one being entailing the improvement or the extermination of others; it follows, that the amount of organic change in the fossils of consecutive formations probably serves as a fair measure of the lapse of actual time. A number of species, however, keeping in a body might remain for a long period unchanged, whilst within this same period, several of these species, by migrating into new countries and coming into competition with foreign associates, might become modified; so that we must not overrate the accuracy of organic change as a measure of time. During early periods of the earth's history, when the forms of life were probably fewer and simpler, the rate of change was probably slower; and at the first dawn of life, when very few forms of the simplest structure existed, the rate of change may have been slow in an extreme degree. The whole history of the world, as at present known, although of a length quite incomprehensible by us, will hereafter be recognised as a mere fragment of time, compared with the ages which have elapsed since the first creature, the progenitor of innumerable extinct and living descendants, was created.

In the distant future I see open fields for far more important researches. Psychology will be based on a new foundation, that of the necessary acquirement of each mental power and capacity by gradation. Light will be thrown on the origin of man and his history.

Authors of the highest eminence seem to be fully satisfied with the view that each species has been independently created. To my mind it accords better with what we know of the laws impressed on matter by the Creator, that the production and extinction of the past and present inhabitants of the world should have been due to secondary causes, like those determining the birth and death of the individual. When I view all beings not as special creations, but as the lineal descendants of some few beings which lived long before the first bed of the Silurian system was deposited, they seem to me to become ennobled. Judging from the past, we may safely infer that not one living species will transmit its unaltered likeness to a distant futurity. And of the species now living very few will transmit progeny of any kind to a far distant futurity; for the manner in which all organic beings are grouped, shows that the greater number of species of each genus, and all the species of many genera, have left no descendants, but have become utterly extinct. We can so far take a prophetic glance into futurity as to foretel[l] that it will be the common and widely-spread species, belonging to the larger and dominant groups, which will ultimately prevail and procreate new and dominant species. As all the living forms of life are the lineal descendants of those which lived long before the Silurian epoch, we may feel certain that the ordinary succession by generation has never once been broken, and that no cataclysm has desolated the whole world. Hence we may look with some confidence to a secure future of equally inappreciable length. And as natural selection works solely by and for the good of each being, all corporeal and mental endowments will tend to progress towards perfection.

It is interesting to contemplate an entangled bank, clothed with many plants of many kinds, with birds singing on the bushes, with various insects flitting about, and with worms crawling through the damp earth, and to reflect that these elaborately constructed forms, so different from each other, and dependent on each other in so complex a manner, have all been produced by laws acting around us. These laws, taken in the largest sense, being Growth with Reproduction; Inheritance which is almost implied by reproduction; Variability from the indirect and direct action of the external conditions of life, and from use and disuse; a Ratio of Increase so high as to lead to a Struggle for Life, and as a consequence to Natural Selection, entailing Divergence of Character and the Extinction of less-improved forms. Thus, from the war of nature, from famine and death, the most exalted object which we are capable of conceiving, namely, the production of the higher animals, directly follows. There is grandeur in this view of life with its several powers, having been originally breathed into a few forms or into one; and that, whilst this planet has gone cycling on according to the fixed law of gravity, from so simple a beginning endless forms most beautiful and most wonderful have been, and are being, evolved.